Lecture Notes in Bioinformatics   8626

Subseries of Lecture Notes in Computer Science

Matteo Comin   Lukas Käll
Elena Marchiori   Alioune Ngom
Jagath Rajapakse (Eds.)

Pattern Recognition in Bioinformatics

9th IAPR International Conference, PRIB 2014
Stockholm, Sweden, August 21-23, 2014
Proceedings

 Springer

Volume Editors

Matteo Comin
University of Padova
Dept. of Information Engineering
Padova, Italy
E-mail: comin@dei.unipd.it

Lukas Käll
Royal Institute of Technology
School of Biotechnology
Science for Life Laboratory and
Swedish e-Science Research Centre
Solna, Sweden
E-mail: lukas.kall@scilifelab.se

Elena Marchiori
Radboud University
Faculty of Sciences
Dept. of Computer Science
Nijmegen, The Netherlands
E-mail: elenam@cs.ru.nl

Alioune Ngom
University of Windsor
School of Computer Science
Windsor, ON, Canada
E-mail: angom@cs.uwindsor.ca

Jagath Rajapakse
Nanyang Technological University
School of Computer Engineering
Singapore, Singapore
E-mail: asjagath@ntu.edu.sg

ISSN 0302-9743   e-ISSN 1611-3349
ISBN 978-3-319-09191-4   e-ISBN 978-3-319-09192-1
DOI 10.1007/978-3-319-09192-1
Springer Cham Heidelberg New York Dordrecht London

Library of Congress Control Number: 2014943103

LNCS Sublibrary: SL 8 – Bioinformatics

Typesetting: Camera-ready by author, data conversion by Scientific Publishing Services, Chennai, India

Printed on acid-free paper

Springer is part of Springer Science+Business Media (www.springer.com)

Preface

In the post-genomic era, a holistic understanding of biological systems and processes, in all their complexity, is critical in comprehending nature's choreography of life. As a result, bioinformatics involving its two main disciplines, namely, the life sciences and the computational sciences, is fast becoming a very promising multidisciplinary field of research. With the ever-increasing application of large-scale high-throughput technologies, such as gene or protein microarrays and mass spectrometry, and massive parallel sequencing the enormous body of information is growing rapidly. Bioinformaticians are posed with a large number of difficult problems to solve, arising not only due to the complexities in acquiring the molecular information but also due to the size and nature of the generated data sets and/or the limitations of the algorithms required for analyzing these data. The recent advancements in computational and information-theoretic techniques are enabling us to conduct various in silico testing and screening of many lab-based experiments before these are actually performed in vitro or in vivo. These in silico investigations are providing new insights for interpreting and establishing new directions for a deeper understanding. Among the various advanced computational methods currently being applied to such studies, the pattern recognition techniques are mostly found to be at the core of the whole discovery process for apprehending the underlying biological knowledge. Thus, we can safely surmise that the ongoing bioinformatics revolution may, in future, inevitably play a major role in many aspects of medical practice and/or the discipline of life sciences. The aim of the conference on Pattern Recognition in Bioinformatics (PRIB) is to provide an opportunity to academics, researchers, scientists, and industry professionals to present their latest research in pattern recognition and computational intelligence-based techniques applied to problems in bioinformatics and computational biology. It also provides them with an excellent forum to interact with each other and share experiences. The conference is organized jointly by the Royal Institute of Technology - KTH, and IAPR (International Association for Pattern Recognition) Bioinformatics Technical Committee (TC-20). This volume presents the proceedings of the 9th IAPR International Conference on Pattern Recognition in Bioinformatics (PRIB 2014), held in Stockholm (Sweden), August 21–23, 2014. It includes 18 technical contributions that were selected by the international Program Committee. Each of these rigorously reviewed papers was presented orally at PRIB 2014. The proceedings consists of two parts: nine full papers and nine short abstracts.

Many have contributed directly or indirectly toward the organization and success of the PRIB 2014 conference. We would like to thank all the individuals and institutions, especially the authors for submitting the papers and the sponsors for generously providing financial support for the conference. We are very grateful to the Bioinformatics Infrastructure for Life Sciences (BILS) for the sponsorship. Our gratitude goes to the Royal Institute of Technology and IAPR Bioinformatics Technical Committee (TC-20) for supporting the conference in many ways. We would like to express our gratitude to all PRIB 2014 international Program Committee members for their objective and thorough reviews of the submitted papers. We fully appreciate the PRIB 2014 Organizing Committee for their time, efforts, and excellent work. We would also like to thank the Science for Life Laboratory for hosting the symposium and providing technical support. Last, but not least, we wish to convey our sincere thanks to Springer for providing excellent professional support in preparing this volume.

August 2014

Matteo Comin
Lucas Käll
Elena Marchiori
Alioune Ngom
Jagath C. Rajapakse

PRIB 2014 Organization

General Chair

Lucas Käll   KTH, Sweden

Program Co-chairs

Alioune Ngom   University of Windsor, Canada
Jagath C. Rajapakse   Nanyang Technological University, Singapore

Publicity Chair

Elena Marchiori   Radboud University of Nijmegen,
                                   The Netherlands

Publication Chair

Matteo Comin   University of Padova, Italy

PRIB Steering Committee

Raj Acharya   Penn State, USA
Shandar Ahmad   Tokyo Tech NIBIO, Japan
Madhu Chetty Monash   University, Australia
Tom Heskes Radboud   University of Nijmegen, The Netherlands
Visakan Kadirkamanathan   University of Sheffield, UK
Elena Marchiori   Radboud University of Nijmegen,
                                   The Netherlands
Jagath C. Rajapakse   Nanyang Technological University, Singapore
Marcel Reinders   Delft University of Technology,
                                   The Netherlands
Dick de Ridder   Delft University of Technology,
                                   The Netherlands
Guido Sanguinetti   University of Edinburgh, UK
Jun Sese   Tokyo Tech NIBIO, Japan

Program Committee

Raj Acharya   Pennsylvania State University, USA
Shandar Ahmad   National Institute of Biomedical Innovation,
                                   Osaka
Tatsuya Akutsu   Kyoto University, Japan

Sahar Al Seesi	University of Connecticut, USA
Pedro Ballester	Cambridge University, UK
Justin Bedo	NICTA, Australia
Chengpeng Bi	Children's Mercy Hospitals, University of Missouri, USA
Mikael Boden	The University of Queensland, Australia
Vladimir Brusic	Dana-Farber Cancer Institute, Harvard Medical School, USA
Conrad Burden	Australian National University, Australia
Keith C.C. Chan	The Hong Kong Polytechnic University, China
Kuo-Sheng Cheng	National Cheng Kung University, Taiwan
Francis Chin	The University of Hong Kong, China
Sung-Bae Cho	Yonsei University, Korea
Dominique Chu	University of Kent, UK
Haibin Duan	Beijing University of Aeronautics and Astronautics, China
Beatrice Duval	LERIA, France
Mansour Ebrahimi	University of Qom, Iran
Esmaeil Ebrahimie	Shiraz University, Iran
Michael Fernandez	KIT, Canada
Christoph M. Friedrich	University of Applied Science and Arts Dortmund, Germany
Kazuhiro Fukui	University of Tsukuba, Japan
Michael Gromiha	IIT Madras, India
Yann Guermeur	CNRS, France
Xiaoxu Han	University of Iowa, USA
Jin-Kao Hao	University of Angers, France
Morihiro Hayashida	Institute for Chemical Research, Kyoto University, Japan
David Hecht	Southwestern College, USA
Md. Tamjidul Hoque	University of New Orleans, USA
Liang-Tsung Huang	Mingdao University, Taiwan
Xiuzhen Huang	Arkansas State University, USA
Zina M. Ibrahim	University of Windsor, UK
Seiya Imoto	University of Tokyo, Japan
Takashi Ishida	Tokyo Institute of Technology, Japan
R. Krishna Murthy Karuturi	Genome Institute of Singapore, Singapore
Marta Kasprzak	Poznan University of Technology, Poland
Yuki Kato	Nara Institute of Science and Technology, Japan
Amit Kumar	Institute of Technology Hauz Khas, New Delhi, India
Zoe Lacroix	Arizona State University, Japan
En-Shiun Annie Lee	University of Waterloo, Canada
Yifeng Li	University of British Columbia, Canada
Weiguo Liu	Shandong University, China

Raunaq Malhotra	Pennsylvania State University, USA
Marcelo Maraschin	Federal University of Santa Catarina, Brazil
Yuri Matsuzaki	Tokyo Institute of Technology, Japan
Martin Middendorf	University of Leipzig, Germany
Mariofanna Milanova	University of Arkansas, USA
Vadim Mottl	Russian Academy of Sciences, Russia
Julio Cesar Nievola	Pontifícia Universidade Católica do Paraná, Brazil
Johan Nystrom-Persson	The University of Tokyo, Japan
Masahito Ohue	Tokyo Tech, Japan
Diana Oliveira	Universidade Estadual do Ceará, Brazil
Helen Piontkivska	Kent State University, UK
Kiran Sree Pokkuluri	BVCEC, India
Gianfranco Politano	Politecnico di Torino, Italy
Dusan Popovic	Katholieke Universiteit Leuven, Belgium
Miguel Rocha	University of Minho, Portugal
Juan J. Rodriguez	University of Burgos, Spain
Dario Rojas	Universidad Católica del Norte, Chile
Luis Rueda	University of Windsor, Canada
Hiroto Saigo	Kyushu Institute of Technology, Japan
Saras Saraswathi	Iowa State University, USA
Bertil Schmidt	University of Mainz, Germany
Christian Schönbach	Nazarbayev University, Kazakhstan
Muhammad Shoaib Sehgal	University of Queensland, Japan
Oleg Seredin	Tula State University, Russia
Ugur Sezerman	Sabanci University, Turkey
Boris Shabash	Simon Fraser University, Canada
Dharmendra Sharma	University of Canberra, Australia
Alexandros Stamatakis	Technical University of Munich, Germany
Bela Stantic	Griffith University, Australia
Gregor Stiglic	University of Maribor, Slovenia
Valentina Sulimova	Tula State University, Russia
Chendra Hadi Suryanto	University of Tsukuba, Japan
Marta Szachniuk	Polish Academy of Sciences, Poland
Y-H. Taguchi	Chuo University, Japan
Takeyuki Tamura	Kyoto University, Japan
Qichuan Tian	Beijing University of Civil Engineering and Architecture, China
Ivan Torshin	MIPT, Russia
Herbert Treutlein	Cytopia Research, Australia
Herbert H. Tsang	Trinity Western University, Canada
Chun-Wei Tung	Kaohsiung Medical University, Taiwan

Twan Van Laarhoven	Radboud University, The Netherlands
Junbai Wang	Radium Hospital, Norway
Lusheng Wang	City University of Hong Kong, China
Pengyi Yang	National Institutes of Health (NIH), USA
Xing-Ming Zhao	Tongji University, China

Table of Contents

Full Papers

Acquiring Decision Rules for Predicting Ames-Negative
Hepatocarcinogens Using Chemical-Chemical Interactions   1
   Chun-Wei Tung

Using Topology Information for Protein-Protein Interaction
Prediction   10
   Adriana Birlutiu and Tom Heskes

Biases of Drug–Target Interaction Network Data   23
   Twan van Laarhoven and Elena Marchiori

Logol: Expressive Pattern Matching in Sequences. Application to
Ribosomal Frameshift Modeling   34
   Catherine Belleannée, Olivier Sallou, and Jacques Nicolas

Evolutionary Algorithm Based on New Crossover for the Biclustering
of Gene Expression Data   48
   Ons Maâtouk, Wassim Ayadi, Hend Bouziri, and Beatrice Duval

SFFS-SW: A Feature Selection Algorithm Exploring the Small-World
Properties of GNs   60
   Fábio Fernandes da Rocha Vicente and Fabrício Martins Lopes

CytomicsDB: A Metadata-Based Storage and Retrieval Approach for
High-Throughput Screening Experiments   72
   E. Larios, Y. Zhang, L. Cao, and F.J. Verbeek

CUDAGRN: Parallel Speedup of Inferring Large Gene Regulatory
Networks from Expression Data Using Random Forest   85
   Seyed Ziaeddin Alborzi, D.A.K. Maduranga, Rui Fan,
   Jagath C. Rajapakse, and Jie Zheng

Supervised Selective Kernel Fusion for Membrane Protein Prediction . . .   98
   Alexander Tatarchuk, Valentina Sulimova, Ivan Torshin,
   Vadim Mottl, and David Windridge

Short Abstracts

Analysis of miRNA Expression Profiles in Breast Cancer Using
Biclustering   110
   Antonino Fiannaca, Massimo La Rosa, Laura La Paglia,
   Riccardo Rizzo, and Alfonso Urso

Gram-Positive and Gram-Negative Subcellular Localization Using
Rotation Forest and Physicochemical-Based Features 112
    Abdollah Dehzangi, Rhys Heffernan, James Lyons, Alok Sharma,
    Kuldip Paliwal, and Abdul Sattar

Data Driven Feature Selection for RNA-Seq Differential Expression
Analysis 114
    Henry Han

Intramuscular Fat Percentage Estimation through Ultrasound Images ... 116
    José Luis Nunes, Alicia Fernandez, and Federico Lecumberry

An Integrated Approach of Gene Expression and DNA-methylation
Profiles of WNT Signaling Genes Uncovers Novel Prognostic Markers
in Acute Myeloid Leukemia 123
    Erdogan Taskesen, Frank J.T. Staal, and Marcel J.T. Reinders

Improving Performance of the eXtasy Model by Hierarchical
Sampling 125
    Dusan Popovic, Alejandro Sifrim, Jesse Davis, Yves Moreau, and
    Bart De Moor

Ensemble Neural Networks Scoring Functions for Accurate Binding
Affinity Prediction of Protein-Ligand Complexes 129
    Hossam M. Ashtawy and Nihar R. Mahapatra

Integration of Gene Expression and DNA-methylation Profiles Improves
Molecular Subtype Classification in Acute Myeloid Leukemia 131
    Erdogan Taskesen, Sepideh Babaei, Marcel J.T. Reinders, and
    Jeroen de Ridder

The Relative Vertex Clustering Value – A New Criterion for the Fast
Discovery of Functional Modules in Protein Interaction Networks 133
    Alioune Ngom, Yifeng Li, and Zina M. Ibrahim

Author Index 135

Acquiring Decision Rules for Predicting Ames-Negative Hepatocarcinogens Using Chemical-Chemical Interactions

Chun-Wei Tung

School of Pharmacy, Kaohsiung Medical University, 80708, Taiwan
Ph.D. Program in Toxicology, Kaohsiung Medical University, 80708, Taiwan
National Environmental Health Research Center, National Health Research Institutes,
Miaoli County 35053, Taiwan
cwtung@kmu.edu.tw
http://cwtung.kmu.edu.tw

Abstract. Chemical carcinogenicity is an important safety issue for the evaluation of drugs and environmental pollutants. The Ames test is useful for detecting genotoxic hepatocarcinogens. However, the assessment of Ames-negative hepatocarcinogens depends on 2-year rodent bioassays. Alternative methods are desirable for the efficient identification of Ames-negative hepatocarcinogens. This study proposed a decision tree-based method using chemical-chemical interaction information for predicting hepatocarcinogens. It performs much better than that using molecular descriptors with accuracies of 86% and 76% for validation and independent test, respectively. Four important interacting chemicals with interpretable decision rules were identified and analyzed. With the high prediction performances, the acquired decision rules based on chemical-chemical interactions provide a useful prediction method and better understanding of Ames-negative hepatocarcinogens.

Keywords: Ames-Negative Hepatocarcinogens, Decision Tree, Chemical-Chemical Interaction, Interpretable Rule, Toxicology.

1 Introduction

The assessment of carcinogenicity is crucial for drug development that is based on 2-year rodent bioassays. The bioassays are labor-intensive, time-consuming and expensive. Chemical carcinogens can be classified as either genotoxic (mutagenic) or non-genotoxic (non-mutagenic) agents according to the mechanism of action [1]. Several short-term *in vitro* and *in vivo* assays have been developed to assess genotoxic agents by measuring DNA damage, mutagenic effects, and chromosomal aberrations [2]. Among the assays, the predictivity of Ames test has been extensively studied for carcinogenicity. The Ames test is useful for identifying mutagenic carcinogens with an accuracy of 80% [3,4]. However, 48% of Ames-negative chemicals are carcinogens [5] and additional bioassays do not help in detecting carcinogens from Ames-negative chemicals [6]. It is desirable

M. Comin et al. (Eds.): PRIB 2014, LNBI 8626, pp. 1–9, 2014.

to develop alternative methods for assessing carcinogenicity of Ames-negative chemicals.

A quantitative structure-activity relationship (QSAR) model has been evaluated for its prediction performance of non-genotoxic hepatocarcinogens. However, the accuracy is only slightly better than random (55%) [7]. Recently, toxicogenomics (TGx) correlating gene expression profiles and toxicity endpoints has emerged as important alternative methods. The TGx methods performed well in non-genotoxic hepatocarcinogenicity with a test accuracy of 80% [7,8]. However, gene expression profiles are only available for a small number of chemicals. It is highly expensive to conduct a large-scale TGx study for hepatocarcinogens.

Chemical-protein interaction (CPI) information has been proposed to predict non-genotoxic hepatocarcinogens with a high accuracy of 86% using only one protein biomarker [9]. Notably, both the aforementioned TGx and CPI methods were performed on a small dataset with less than 62 chemicals. Although the CPI information is useful for analyzing and predicting hepatocarcinogens, the information is incomplete that many chemical-protein pairs have not been studied yet. The development of computational methods for a large number of chemicals is desirable.

This study constructed a relatively large dataset consisting of 166 chemicals by extracting information of Ames-negative chemicals and corresponding hepatocarcinogenicity from NCTRlcdb [10]. The more complete chemical-chemical interaction (CCI) information from STITCH database [11] was proposed to predict hepatocarcinogens based on the assumption that interactive chemicals are more likely to share similar toxicity. The CCI information has been successfully applied to predict various chemical activities such as cancer drugs and chemical toxicity [12,13].

In order to acquire rule-based knowledge, interpretable decision tree classifiers were applied to predict hepatocarcinogenicity with accuracies of 85% and 76% for validation and independent test, respectively. The CCI-based method performs much better than a QSAR-based method with 12% and 6% improvements in terms of accuracy for validation and independent test, respectively. The decision rules were also analyzed to give insights into hepatocarcinogenicity.

2   Materials and Methods

2.1   Dataset

Ames-negative rodent hepatocarcinogens and noncarcinogens were extracted from a liver cancer database NCTRlcdb [10]. The annotations of organ-specific carcinogenicity and mutagenicity are available for 999 chemical compounds. Mutagenic chemicals (Ames-positive) were firstly removed. Subsequently, hepatocarcinogens and noncarcinogens were identified according to the field of OVER-ALL. Six noncarcinogens without corresponding chemical-chemical interaction data were also excluded. The final dataset consists of 73 hepatocarcinogens and

93 noncarcinogens. The dataset was randomly divided into three datasets with similar ratios between hepatocarcinogens and noncarcinogens for training (60%), validation (20%) and independent test (20%). The three datasets are available at http://cwtung.kmu.edu.tw/nghc.

2.2 Chemical Descriptors

The software of PaDEL-Descriptor [14] was utilized to generate chemical descriptors from chemical 2D structures extracted from PubChem database. The final feature vector is a 1610-dimensional vector consisting of 770 1D and 2D descriptors and 840 PubChem fingerprints.

2.3 Chemical-Chemical Interactions

Chemical-chemical interaction (CCI) data are obtained from STITCH 3.1 database [11], an aggregated database of interactions connecting over 300,000 chemicals and 2.6 million proteins from 1,133 organisms. For each CCI, there is a combined score calculated by combining four evidence sources of experiments, databases, text-mining and similarity. In this study, the scores divided by 1,000 are utilized to represent CCI features. The scores are ranging from 0 (low confident) to 1 (high confident).

2.4 Decision Tree Algorithm

Decision tree algorithms capable of generating interpretable rules are widely used in various biological problems such as immunogenic peptides [15], ubiquitylation sites [16] and esophageal squamous cell carcinoma [17]. In this study, the decision tree method C5.0 is applied to construct decision tree classifiers and derive interpretable rules based on CCI features for predicting hepatocarcinogenicity. C5.0 is an improved version of C4.5 with smaller trees and less computation time [18]. The implementation of R package C50 is utilized in this study [19].

The construction of a decision tree is briefly described as follows. First, information gain is utilized to rank features. Second, the top-ranking features are iteratively appended as nodes to split data into subsets. The tree growing process stops when the data subset in each leaf node belongs to the same class. The fully-grown tree is prone to over-fit the training data. Therefore, a pruning process is applied to reduce the tree size by replacing a subtree with a leaf node to avoid over-fitting problems. The pruning process is based on a default threshold value of 25% confidence. The samples in the leaf node are the covered samples of the rule. The class label of a leaf node is determined by using a majority rule. The samples with a relative small size in the leaf node are regarded as misclassified samples. The final decision tree can directly generate if-then rules where one leaf node corresponds to one rule.

2.5   Feature Selection

The selection of important features can provide better insights into the biological problems and improve prediction performances [20,16,21,17]. This study utilized a two-step feature selection method. First, features with near zero variances were removed. Baseline models are constructed by using features whose variances are not near zero. Second, a wrapper-based feature selection method using a minimum redundancy-maximum relevancy (mRMR) method [22] is utilized to identify important CCI features for analyses and development of prediction methods. The mRMR selection process is described as follows. First, mRMR is utilized to rank the importance of CCI features. Subsequently, a sequential backward feature elimination algorithm is applied to iteratively remove CCI features with lowest ranks for selecting a subset of CCI features giving the highest 10-fold cross-validation (10-CV) accuracy. The selected feature subset is used to construct a decision tree model for predicting hepatocarcinogens.

2.6   Performance Measurement

To evaluate classifiers for their prediction performance, the widely used 10-fold cross-validation method is applied. Four measurements were used to evaluate prediction performances including sensitivity, specificity, precision and accuracy, defined as follows:

$$\text{Sensitivity} = \frac{TP}{TP + FN}, \tag{1}$$

$$\text{Specificity} = \frac{TN}{TN + FP}, \tag{2}$$

$$\text{Precision} = \frac{TP}{TP + FP}, \cdot \tag{3}$$

$$\text{Accuracy} = \frac{TP + TN}{TP + FP + FN + TN}, \tag{4}$$

where *TP*, *FP*, *FN* and *TN* are the numbers of true positives, false positives, false negatives and true negatives, respectively. In this work, accuracy is used as the major indicator for estimating the performance of classifiers.

3   Results and Discussion

3.1   Selection of Informative Features

A baseline model using all 223 CCI features whose variances are not near zero is firstly evaluated for comparison. The accuracies of 10-CV and validation for the baseline model are 64% and 72.73% using training and validation datasets, respectively. To identify informative features for Ames-negative hepatocarcinogens, the sequential backward feature elimination algorithm was applied to the training dataset consisting of 45 hepatocarcinogens and 55 noncarcinogens.

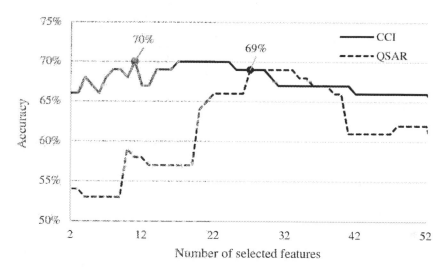

Fig. 1. The cross-validation performance for various numbers of selected features

The feature selection process and corresponding 10-CV accuracies are shown in Figure 1. Based on the training dataset, the algorithm selected a small subset of 11 CCI features giving a highest 10-CV accuracy of 70% that is 6% higher than the baseline model. The feature selected model performs well in training dataset with an accuracy of 82%. To evaluate the validation performance of the feature selected model, a decision tree model constructed by using the 11 CCI features and training dataset was utilized to classify chemicals in the validation dataset consisting of 14 hepatocarcinogens and 19 noncarcinogens. A high validation accuracy of 84.85% is obtained from the feature selected model that is 12% higher than the baseline model. Detailed performance is shown in Table 1. In addition to the mRMR method, three additional methods of chi-square test, variable importance of random forest, and relief were also evaluated with worse validation accuracies of 72.73%, 69.70% and 69.70%, respectively. The mRMR method aiming to select a feature subset of minimum redundancy and maximum relevancy might be able to avoid overfitting problems.

3.2 Independent Test Performance

To further evaluate the prediction performance of the proposed method, the decision tree model constructed by using the 11 selected CCI features was utilized to predict the chemicals in the independent test dataset consisting of 14 hepatocarcinogens and 19 noncarcinogens. The test performances are 75.76%, 50.00%, 94.74% and 87.50% for accuracy, sensitivity, specificity and precision, respectively. Compared to the test accuracy of the baseline model (66.67%), the constructed decision tree model performs well with 9% improvement. The CCI-based

Table 1. Prediction performance

	Validation		Test	
Method	CCI	QSAR	CCI	QSAR
Accuracy	84.85%	72.73%	75.76%	69.70%
Sensitivity	78.57%	57.14%	50.00%	71.43%
Specificity	89.47%	84.21%	94.74%	68.42%
Precision	84.62%	72.73%	87.50%	62.50%
AUC	0.8421	0.7030	0.7180	0.6880

model with high performances of precision and specificity is expected to be a useful tool for screening Ames-negative hepatocarcinogens. The detailed performance of the constructed model is shown in Table 1.

3.3 Comparison to Quantitative Structure-Activity Relationship (QSAR) Models

For comparison, a QSAR model was developed using the same feature selection algorithm and decision tree classifier. After feature selection, the QSAR model with a 10-CV accuracy of 69% is slightly worse than the CCI-based model (Figure 1). As shown in Table 1, the QSAR model with 27 selected features performs much worse than the CCI-based model in both validation and test dataset. The prediction accuracies of the CCI-based model are 12% and 6% higher than that of the QSAR model for validation and independent test, respectively. Due to different specificity levels of CCI-based and QSAR models, it is hard to conclude the superiority of the CCI-based model. An additional nonparametric measurement of area under receiver operating characteristic (ROC) curve (AUC) is applied to evaluate the CCI-based and QSAR models. As shown in Table 1, results show that the CCI-based model is better than the QSAR model with 14% and 3% improvement on validation and test datasets, respectively.

3.4 Decision Rules for Ames-Negative Hepatocarcinogenicity

To better understand the relationship between important CCI features and Ames-Negative Hepatocarcinogenicity, the decision tree model constructed by using the training dataset and 11 selected CCI features is shown in Figure 2. Five decision rules corresponding to five leaf nodes can be derived from the decision tree. In brief, a chemical interacting with one of the four chemicals is a hepatocarcinogen that correctly predict 27 hepatocarcinogens. Otherwise, it is a noncarcinogen that 55 noncarcinogens are correctly predicted with 18 miss-classified hepatocarcinogens. The four compounds are di-(4-aminophenyl)ether (CID000007579), ethane (CID000006324), 2-acetylaminofluorene (CID000005897), and deoxyguanosine (CID000187790). Among the four compounds, the di-(4-aminophenyl)ether and 2-acetylaminofluorene are Ames-positive carcinogens.

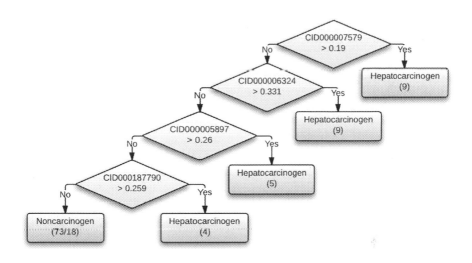

Fig. 2. Decision tree classifier for Ames-negative hepatocarcinogens

4   Conclusions

The development of computational methods for the assessment of hepatocarcino-genicity is important for efficient drug development compared to the traditional 2-year rodent bioassays. Most of the non-mutagenic hepatocarcinogens could be identified by the *in vitro* Ames test. However, it is desirable to develop alternative methods for assessing Ames-negative hepatocarcinogens. The acquisition of rules for efficient recognition of Ames-negative hepatocarcinogens is especially important for practical application. This study proposed a decision tree-based method using the CCI information and mRMR feature selection method for the acquisition of decision rules for predicting hepatocarcinogenicity of Ames-negative chemicals. The prediction model performs well with validation and test accuracies of 85% and 76%, respectively. The acquired simple decision rules are useful for identifying Ames-negative hepatocarcinogens with high specificity and precision. Future works can be the application and comparison of other machine learning methods to improve the prediction performance of Ames-negative hepatocarcinogens.

Acknowledgement. The author would like to acknowledge the financial support from National Science Council of Taiwan (NSC 101-2311-B-037-001-MY2), Kaohsiung Medical University Research Foundation (KMU-M103009), NSYSU-KMU Joint Research Project (NSYSUKMU103-P002), and National Health Research Institutes (EH-103-PP-09).

References

1. Hayashi, Y.: Overview of genotoxic carcinogens and non-genotoxic carcinogens. Exp. Toxicol. Pathol. 44, 465–471 (1992)
2. Weisburger, J.H., Williams, G.M.: The distinction between genotoxic and epigenetic carcinogens and implication for cancer risk. Toxicol. Sci. 57, 4–5 (2000)
3. Zeiger, E.: Identification of rodent carcinogens and noncarcinogens using genetic toxicity tests: premises, promises, and performance. Regul. Toxicol. Pharmacol. 28, 85–95 (1998)
4. Benigni, R., Bossa, C., Tcheremenskaia, O., Giuliani, A.: Alternatives to the carcinogenicity bioassay: in silico methods, and the in vitro and in vivo mutagenicity assays. Expert Opin. Drug Metab. Toxicol. 6, 809–819 (2010)
5. Cunningham, A.R., Carrasquer, C.A., Qamar, S., Maguire, J.M., Cunningham, S.L., Trent, J.O.: Global structure-activity relationship model for nonmutagenic carcinogens using virtual ligand-protein interactions as model descriptors. Carcinogenesis 33, 1940–1945 (2012)
6. Zeiger, E.: Historical perspective on the development of the genetic toxicity test battery in the united states. Environ. Mol. Mutagen. 51, 781–791 (2010)
7. Liu, Z., Kelly, R., Fang, H., Ding, D., Tong, W.: Comparative analysis of predictive models for nongenotoxic hepatocarcinogenicity using both toxicogenomics and quantitative structure-activity relationships. Chem. Res. Toxicol. 24, 1062–1070 (2011)
8. Yamada, F., Sumida, K., Uehara, T., Morikawa, Y., Yamada, H., Urushidani, T., Ohno, Y.: Toxicogenomics discrimination of potential hepatocarcinogenicity of nongenotoxic compounds in rat liver. J. Appl. Toxicol. (2012)
9. Tung, C.W.: Prediction of non-genotoxic hepatocarcinogenicity using chemical-protein interactions. In: Ngom, A., Formenti, E., Hao, J.-K., Zhao, X.-M., van Laarhoven, T. (eds.) PRIB 2013. LNCS, vol. 7986, pp. 231–241. Springer, Heidelberg (2013)
10. Young, J., Tong, W., Fang, H., Xie, Q., Pearce, B., Hashemi, R., Beger, R., Cheeseman, M., Chen, J., Chang, Y.C., Kodell, R.: Building an organ-specific carcinogenic database for sar analyses. J. Toxicol. Environ. Health A 67, 1363–1389 (2004)
11. Kuhn, M., Szklarczyk, D., Franceschini, A., von Mering, C., Jensen, L.J., Bork, P.: Stitch 3: zooming in on protein-chemical interactions. Nucleic Acids Res. 40, D876–D880 (2012)
12. Lu, J., Huang, G., Li, H.P., Feng, K.Y., Chen, L., Zheng, M.Y., Cai, Y.D.: Prediction of cancer drugs by chemical-chemical interactions. PLoS One 9, e87791 (2014)
13. Chen, L., Lu, J., Luo, X., Feng, K.Y.: Prediction of drug target groups based on chemical-chemical similarities and chemical-chemical/protein connections. Biochim. Biophys. Acta 1844, 207–213 (2014)
14. Yap, C.W.: Padel-descriptor: an open source software to calculate molecular descriptors and fingerprints. J. Comput. Chem. 32, 1466–1474 (2011)
15. Tung, C.W., Ziehm, M., Kämper, A., Kohlbacher, O., Ho, S.Y.: Popisk: T-cell reactivity prediction using support vector machines and string kernels. BMC Bioinformatics 12, 446 (2011)
16. Tung, C.W., Ho, S.Y.: Computational identification of ubiquitylation sites from protein sequences. BMC Bioinformatics 9, 310 (2008)

17. Tung, C.W., Wu, M.T., Chen, Y.K., Wu, C.C., Chen, W.C., Li, H.P., Chou, S.H., Wu, D.C., Wu, I.C.: Identification of biomarkers for esophageal squamous cell carcinoma using feature selection and decision tree methods. The Sci. World J. 2013, 782031 (2013)
18. Quinlan, J.: C4. 5: programs for machine learning (1993)
19. Kuhn, M., Weston, S.. Code for C5.0 by R. Quinlan, N.C.C.: C50: C5.0 Decision Trees and Rule-Based Models (2014); R package version 0.1.0-016
20. Tung, C.W.: Prediction of pupylation sites using the composition of k-spaced amino acid pairs. J. Theor. Biol. 336, 11–17 (2013)
21. Tung, C.W., Ho, S.Y.: Popi: predicting immunogenicity of mhc class i binding peptides by mining informative physicochemical properties. Bioinformatics 23, 942–949 (2007)
22. De Jay, N., Papillon-Cavanagh, S., Olsen, C., El-Hachem, N., Bontempi, G., Haibe-Kains, B.: Mrmre: an r package for parallelized mrmr ensemble feature selection. Bioinformatics 29, 2365–2368 (2013)

Using Topology Information
for Protein-Protein Interaction Prediction

Adriana Birlutiu[1] and Tom Heskes[2]

[1] Institute for Computing and Information Sciences,
Radboud University Nijmegen, The Netherlands and Faculty of Science,
"1 Decembrie 1918" University, Alba-Iulia, Romania
adrianab@cs.ru.nl
[2] Institute for Computing and Information Sciences,
Radboud University Nijmegen, The Netherlands
tomh@cs.ru.nl

Abstract. The reconstruction of protein-protein interaction networks is nowadays an important challenge in systems biology. Computational approaches can address this problem by complementing high-throughput technologies and by helping and guiding biologists in designing new laboratory experiments. The proteins and the interactions between them form a network, which has been shown to possess several topological properties. In addition to information about proteins and interactions between them, knowledge about the topological properties of these networks can be used to learn accurate models for predicting unknown protein-protein interactions. This paper presents a principled way, based on Bayesian inference, for combining network topology information jointly with information about proteins and interactions between them. The goal of this combination is to build accurate models for predicting protein-protein interactions. We define a random graph model for generating networks with topology similar to the ones observed in protein-protein interaction networks. We define a probability model for protein features given the absence/presence of an interaction and combine this with the random graph model by using Bayes' rule, to finally arrive at a model incorporating both topological and feature information.

Keywords: protein-protein interaction, Bayesian methods, network analysis.

1 Introduction

Knowledge about protein-protein interactions (PPIs) is essential to the understanding of the cellular functions and biological processes inside a living cell. Deciphering the entire network of PPIs of an organism is a very complex task since these interactions can only be established by costly and tedious laboratory experiments. Computational techniques for predicting PPIs have become standard tools to address this problem, complementing their experimental counterparts. Accurately predicting which proteins might interact can help in designing and guiding future laboratory experiments. Therefore, developing computational

M. Comin et al. (Eds.): PRIB 2014, LNBI 8626, pp. 10–22, 2014.

methods that can accurately predict PPIs is currently an active research area. A number of computational approaches for PPI prediction have been developed over the years. These methods differ in feature information used for PPI prediction, for example genomic data, phylogenetic trees.

A recent trend in computational approaches for predicting PPIs is to frame this problem in a supervised learning setting. That is, information about proteins and labels for protein pairs as interacting or not, supervise the estimation of a function that can predict whether an interaction exists or not between two proteins. PPI prediction can thus be seen as a pattern recognition problem, i.e., find patterns in the interacting protein pairs that do not exist in the non-interacting pairs. This can be further framed as a binary classification problem which takes as input a set of features for a protein pair and gives as output a label: interact or non-interact. Binary classification has been studied extensively in the machine learning community, and many algorithms designed to solve it have been also applied for predicting PPIs, including Bayesian networks [9], kernel-based methods [1,31], logistic regression [14,27], SVMs [26] decision trees and random forest based methods [33,22,2], metric or kernel learning [31] and [7,6,5]. Very recently, other machine learning paradigms, such as, active learning, multi-task learning, and semi-supervised learning, have also been employed for improving the prediction of PPIs [18,24,11].

In addition to information about proteins and interactions between them, PPI networks are characterized by several topological properties [10,15,4,21,28]. Network topology can uncover important biological information that is independent of other available biological information [25,13]. One of the most important topological properties is the existence of a few nodes in the networks, called hubs, which have many links with the other nodes, while most of the nodes have just a few links. This characteristic is present in PPI networks and also in other real-world networks, such as the internet and citation networks. Topology only has been shown to be able to predict protein functions [17] and PPIs [12] and to complement sequence information in various biological tasks, like for example, homology detection [16]. Summarizing, we can distinguish two types of information that can be used for predicting PPIs: first, information about proteins and labels for protein pairs as interacting or not, and second, information about topological properties of PPI networks. These two sources of information can complement each other and are both valuable for constructing models which can accurately predict interactions between proteins.

In this contribution, we present a principled way of combining topology and feature information for constructing models for predicting PPIs. We combine models that have been previously used for modeling each type of information separately. We use a random graph generator for addressing the topology information and a naive Bayes model for addressing the feature information. We show that by making a few simplifying assumptions, both topological and protein information can be incorporated and we show experimentally that this improves the prediction accuracy in two PPI networks.

2   Models and Methods

The approach that we use to combine topology and feature information is graphically summarized in Figure 1. It consists of a random graph generator model and a naive Bayes model which are combined using Bayes' rule to finally arrive to a logistic regression model (we will ignore for the moment the details of this figure but come back to it throughout the section). The random graph

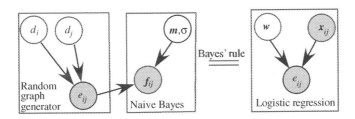

Fig. 1. Graphical representation of the model which combines topology and feature information. Left box: random graph generator model. Center box: naive Bayes model. Right box: the result of applying Bayes' rule, the model which combines topology and feature information.

generator gives rise to networks which based on topology can all be plausible hypotheses for the PPI network that we want to reconstruct. Incorporating the actual data will reduce this set of plausible hypotheses to just a few, out of which we can pick the one which has the highest likelihood. We implement this in a Bayesian framework by treating our random graph model as a prior and define a probability model for the features given the absence/presence of an edge and combine these two using Bayes' rule, to finally arrive at a model incorporating both topological and feature information. The way in which each of these models is constructed and then combined is detailed in the rest of this section.

2.1   Topological Properties of PPI Networks

We will focus on one essential topological characteristics of PPI networks: the node degree distribution. The degree of a node represents the number of connections the node has with the other nodes in the network. The probability distribution of these degrees over the whole network, $p(k)$, is defined as the fraction of nodes in the network with degree k,

$$p(k) = \frac{N_k}{N} ,$$

where N is the total number of nodes in the network and N_k is the number of nodes with degree k. The majority of real-world networks have a node degree distribution that is highly right-skewed, which means that most of the nodes have low degrees, while a small number of nodes, known as "hubs", have high degrees. The degree of hubs is typically several order of magnitudes larger than the average degree of a node in the network.

2.2   Random Graph Generator

The first step of our approach is to define a model for generating networks with the node degree distribution similar to the one of PPI networks (the left-hand side box of Figure 1). The random graph generator that we define here is inspired by the general random graph method [3]. The general random graph method assigns each node with its expected degree and edges are inserted according to a probability proportional to the product of the degrees of the two endpoints, i.e., the probability of an edge between two nodes i and j is proportional to the product of the expected degrees of the nodes i and j. We introduce a latent variable, d_i, related to the degree of node i, i.e., d_i is roughly proportional to the degree of node i. Let e_{ij} be a random variable with two possible values: $e_{ij} = 1$ if a link is present between nodes i and j, and $e_{ij} = -1$ if there is no link. In Figure 1, the random variables d_i and d_j are represented by white color circles because they are unobserved while e_{ij} is represented by a gray color circle because it is observed.

Our model generates links in the network as follows,

$$p(e_{ij}|d_i, d_j) \propto (\sqrt{d_i d_j})^{e_{ij}} = \exp\left[e_{ij}\frac{1}{2}(\log d_i + \log d_j)\right], \qquad (1)$$

$$p(e_{ij} = 1|d_i, d_j) \propto \sqrt{d_i d_j}$$
$$p(e_{ij} = -1|d_i, d_j) \propto \frac{1}{\sqrt{d_i d_j}}$$

$$p(e_{ij} = 1|d_i, d_j) = \frac{p(e_{ij}=1|d_i,d_j)}{p(e_{ij}-1|d_i,d_j)+p(e_{ij}=-1|d_i,d_j)}$$
$$= \frac{\sqrt{d_i d_j}}{\sqrt{d_i d_j}+\frac{1}{\sqrt{d_i d_j}}} = \frac{d_i d_j}{1+d_i d_j},$$

$$p(e_{ij} = -1|d_i, d_j) = \frac{p(e_{ij}=-1|d_i,d_j)}{p(e_{ij}=1|d_i,d_j)+p(e_{ij}=-1|d_i,d_j)}$$
$$= \frac{\frac{1}{\sqrt{d_i d_j}}}{\sqrt{d_i d_j}+\frac{1}{\sqrt{d_i d_j}}} = \frac{1}{1+d_i d_j},$$

In order to generate networks with a desired topology and for computational reasons which will become clear later, we consider a log-normal distribution for d_i,

$$p(\log d_i) = \mathcal{N}(\log d_i; m_0, \sigma_0^2), \qquad (2)$$

where m_0 is a scaling parameter, and the parameter σ_0 controls the shape of the distribution. These parameters can be fit such that the networks randomly generated with the model from Equation (1) have the desired topology. We have defined d_i to be roughly proportional to the degree of node i, thus a log-normal distribution for d_i results in a distribution for the degree of node i which is approximately log-normal, which is similar to what is observed in practice.

In summary, the random graph generator for a given topology performs the following steps.

1. Choose m_0 and σ_0 the parameters of the log-normal distribution for d_i.
2. Draw from this distribution a random sample $(d_1, \ldots, d_N)$ of size N the number of nodes in the network.
3. Based on this sample construct the network by inserting edges with probability given in Equation (1).

2.3   Bayesian Framework for Combining Topology and Feature Information

In order to combine the topology and feature information, we treat the random graph model as a prior and define a probability model for the protein pairs features given the absence/presence of an interaction. We make use of a a naive Bayes model to express the likelihood of a protein pairs feature given the absence/presence of an interaction. The likelihood is thus computed as a product of 1-dimensional Gaussian distributions, each Gaussian distribution expressing the probability of a feature component f_{ij}^k given the edge variable e_{ij} and the parameters mean m_k and variance σ,

$$p(\boldsymbol{f}_{ij}|e_{ij}, \boldsymbol{m}, \sigma) = \prod_{k=1}^{D} \mathcal{N}(f_{ij}^k; m_k e_{ij}, \sigma) \propto \prod_{k=1}^{D} \exp\left(-\frac{(f_{ij}^k - e_{ij} m_k)^2}{2\sigma^2}\right). \quad (3)$$

We refer to the center box of Figure 1 for a graphical representation of this model. The naive Bayes model defined above treats the features as independent, which might not be the case in practice. Despite this simplifying assumption, the naive Bayes model is known to be a competitive classification method, with similar performance as the closely related logistic regression algorithm.

The posterior distribution for e_{ij} which combines topology and feature information is computed using Bayes' rule as the product between the prior defined in Equation (1) and the likelihood terms defined in Equation (3), i.e.,

$$p(e_{ij}|\boldsymbol{f}_{ij}, d_i, d_j) \propto p(e_{ij}|d_i, d_j) p(\boldsymbol{f}_{ij}|e_{ij}, d_i, d_j)$$

$$\propto \exp\left(e_{ij}\frac{1}{2}(\log d_i + \log d_j) - \frac{\sum_k (f_{ij}^k - e_{ij} m_k)^2}{2\sigma^2}\right) \quad (4)$$

$$\propto \exp\left(e_{ij}\frac{1}{2}(\log d_i + \log d_j) + \frac{e_{ij}\sum_k f_{ij}^k m_k}{\sigma^2}\right) \quad (5)$$

$$\propto \exp\left(e_{ij}(\sum_{k=1}^{D} \frac{f_{ij}^k m_k}{\sigma^2} + \frac{1}{2}\log d_i + \frac{1}{2}\log d_j)\right) \quad (6)$$

where when going from (4) to (5) we discarded the square terms. In the above, we can ignore any term that does not depend on e_{ij}, since it will only affect the normalization. This includes the term $e_{ij}^2 m_k^2 / \sigma^2$, since $e_{ij} \in \{-1, 1\}$. The normalization term does play a role and, when incorporated, leads to Equation (8) below. The unknown quantities of our model are $\frac{m_k}{\sigma^2}$, $k = \{1, \ldots, D\}$ and $\log d_i$, $i = \{1, \ldots, N\}$, and these will be estimated based on the available training data in a learning procedure that we describe below.

The first step is to adjoin the unknown quantities in a single random variable, that is

$$\boldsymbol{w} = [\frac{m_1}{\sigma^2}, \ldots, \frac{m_D}{\sigma^2}, \frac{1}{2} \log d_1, \ldots, \frac{1}{2} \log d_N] , \tag{7}$$

and the same for the information available, that is protein features and topological information

$$\boldsymbol{x}_{ij} = [\boldsymbol{f}_{ij}, \boldsymbol{t}_{ij}] ,$$

where $\boldsymbol{t}_{ij}$ is the position vector having 1 on positions i and j and 0 everywhere else. Then, the normalized probability that there is an interaction between the proteins i and j from Equation (6) can be rewritten as

$$p(e_{ij}|\boldsymbol{x}_{ij}, \boldsymbol{w}) = \frac{1}{1 + \exp(-2e_{ij}\boldsymbol{w}^T \boldsymbol{x}_{ij})} . \tag{8}$$

Note that in the sum

$$\boldsymbol{w}^T \boldsymbol{x}_{ij} = \sum_{k=1}^{D} w^k f_{ij}^k + \sum_{k=1}^{N} w^{D+k} t_{ij}^k , \tag{9}$$

the first term on the right hand side originates from the protein features information and the second term from the topological information.

The unknown parameter $\boldsymbol{w}$ is learned in a Bayesian framework which consists in setting a prior distribution for it, and updating this prior based on observations. The update is performed using Bayes' rule given below

$$p(\boldsymbol{w}|\text{observations}) \propto \prod_{o=1}^{n_{\text{obs}}} p(e_{ij}^o|\boldsymbol{x}_{ij}^o, \boldsymbol{w}) p(\boldsymbol{w}) . \tag{10}$$

where n_{obs} is the size of the training data, i.e., the number of known interacting/non-interacting protein pairs, and $p(e_{ij}^o|\boldsymbol{x}_{ij}^o, \boldsymbol{w})$ is given in Equation (8). $p(\boldsymbol{w})$ is the prior and we choose it to be a Gaussian distribution

$$p(\boldsymbol{w}) = \mathcal{N}(\boldsymbol{w}; \boldsymbol{\mu}, \boldsymbol{\Sigma}) .$$

The hyperparameters $\boldsymbol{\mu}$ and $\boldsymbol{\Sigma}$ of the prior are chosen such that the topological information is included in the model. This is implemented by making the correspondence with the prior for the latent variables d_i. Recall from Equation (7) that $w^{i+D} = \frac{1}{2} \log d_i$, $i = 1, \ldots, N$ and from Equation (2) that $\log d_i$ is normally distributed, consequently w^{i+D} will also be normally distributed, i.e.,

$$w^{i+D} \sim \mathcal{N}\left(\frac{m_0}{2}, \frac{\sigma_0^2}{4}\right) , i = 1, \ldots, N .$$

The vectors $\boldsymbol{x}_{ij}$ are sparse because their components $\boldsymbol{t}_{ij}$ of dimension N contain only two non-zero elements on positions i and j. This sparsity property can be exploited for making the computations more efficient. Predictions can be done for an unknown interaction between a pair of proteins i', j' characterized by the feature vector $\boldsymbol{x}_{i'j'}$. These predictions can be done either averaging the posterior over $\boldsymbol{w}$ in Equation (8) or by using a point estimate of this posterior, let $\boldsymbol{w}^*$ be the mean of $p(\boldsymbol{w}|\text{observations})$, and computing $p(e_{i'j'}|\boldsymbol{x}_{i'j'}, \boldsymbol{w}^*)$ using Equation (8).

We refer back to the graphical sketch of our model in Figure 1 at the beginning of this section. The box on the left-hand side, corresponds to the random graph generator model. The observation e_{ij}, which expresses the presence or absence of an edge between nodes i and j, depends on the latent variables d_i and d_j which are related to the degrees of nodes i and j. The random graph generator model incorporates feature information through the naive Bayes model with unknown parameters $\boldsymbol{m}$ and σ, represented in the center box. The combination of the two models is obtained using Bayes' rule. The result is shown in the right-hand side box. The unknown quantities d_i, d_j, and $\boldsymbol{m}$, σ are combined in the node $\boldsymbol{w}$ which is unobserved, and $\boldsymbol{f}_{ij}$ together with $\boldsymbol{t}_{ij}$ which is implicitly expressed by indices i and j form the observed quantity $\boldsymbol{x}_{ij}$.

In the experimental evaluation from Section 3 we will compare four models. All the models are based on Equation (10) with a Gaussian prior and likelihood terms of the form given in Equation (8) and they vary in the way of computing the dot product from Equation (9) and on the parameters of the Gaussian prior.

1. Model 1 (Features+Topology): is the model we propose in this work. It makes use of the following dot product

$$\boldsymbol{w}^T \boldsymbol{x}_{ij} = \sum_{k=1}^{D} w^k f_{ij}^k + \sum_{k=1}^{N} w^{D+k} t_{ij}^k \,, \tag{11}$$

and a Gaussian prior with mean $\boldsymbol{\mu}^{1:D} = 0$, $\boldsymbol{\mu}^{D+1:N} = -1.5$ and covariance matrix equal to the identity matrix.

2. Model 2 (Features only): uses only information about proteins, and the dot product is computed as

$$\boldsymbol{w}^T \boldsymbol{x}_{ij} = \sum_{k=1}^{D} w^k f_{ij}^k + w^{D+1} \,. \tag{12}$$

The second term on the right-hand side of Equation (12) is a bias term to address the unbalancedness of the data. This bias term also corresponds to the second term on the right-hand side of Equation (11); for an edge e_{ij} the contributions in Equation (11) are $w^{D+i} + w^{D+j}$ while in Equation (12) we constraint $w^{D+i} = \frac{1}{2} w^{D+1}, \forall i = 1, \ldots, N$. This observation also motivates the choice of the prior for this model: mean $\boldsymbol{\mu}^{1:D} = 0$ and $\mu^{D+1} = -3$ and covariance equal to the identity matrix.

3. Model 3 (Topology only): uses only topology information and the dot product is computed as

$$\boldsymbol{w}^T \boldsymbol{x}_{ij} = \sum_{k=1}^{N} w^k t_{ij}^k \,.$$

The Gaussian prior is of dimension N with mean equal to the vector $\boldsymbol{\mu}^{1:N} = -1.5$ and covariance matrix equal to the identity matrix. The choice for $\boldsymbol{\mu}^{1:N} = -1.5$ corresponds to the log-normal distribution with $m_0 = -3$, thus to a network with a node degree distribution similar to the one observed in PPI networks.

4. Model 4 (Topology-enriched features): uses the information about proteins and about topology in the following form

$$\boldsymbol{w}^T \boldsymbol{x}_{ij} = \sum_{k=1}^{D} w^k f_{ij}^k + w^{D+1} \log(\hat{d}_i + 1) + w^{D+2} \log(\hat{d}_j + 1) \,,$$

where $\hat{d}_i$ and $\hat{d}_j$ are the estimated degrees of nodes i and j computed on the training data. Basically, the features $\boldsymbol{f}_{ij}$ for a pair of proteins i and j are being extended by adding two new columns corresponding to the degrees of nodes i and j computed on the training set. For computational reasons we considered the logarithms of node degrees to which we added 1. The idea behind this model is similar to the one used in [29,24], i.e., the topological features are added to protein features resulting in an enriched set of features. The features are being standardized and the parameters of the Gaussian prior are set to $\boldsymbol{\mu}^{1:D+2} = 0$ and covariance equal to the identity matrix.

3   Results

In this section we discuss the results of the experimental evaluation of the framework proposed here. We compare the performance obtained using information about proteins only, with the performance obtained using topology information only and with the performance obtained with the combination of the two.

3.1   Data Sets

We used two data sets. Details for each of them are given below.

Yeast Data. This data set was borrowed from [5] and it consists of the high confidence physical interactions between proteins highlighted in [30]. The PPI network has 984 nodes (proteins) connected by 2438 links (interactions). We consider all the protein pairs not present in the 2438 interactions as non-interacting. The yeast PPI graph is very sparse, as a result the data is highly unbalanced, with less than 1% from the total examples belonging to the positive class. Each protein has associated a vector of dimension 157 representing gene expression values in various experiments. We constructed the features for protein pairs by summing the individual protein features.

Human Data. This data set was created and made available by [23] and consists of protein pairs with an associated label: interact or non-interact. Each pair of proteins is characterized by a 27-dimensional feature vector. The features were constructed based on Gene Ontology (GO) cell component (1), GO molecular function (1), GO biological process (1), co-occurrence in tissue (1), gene expression (16), sequence similarity (1), homology based (5) and domain interaction (1), where the numbers in brackets correspond to the number of elements contributed by the feature type to the feature vector. Unlike positive interactions, non-interacting pairs are not experimentally reported. Thus, a common strategy is to consider as non-interacting pairs a randomly drawn fraction from the total set of potential protein pairs excluding the pairs known to interact. The resulting data set has 14,608 interacting pairs and 432,197 non-interacting pairs. The PPI graph consists of $24,380$ nodes connected by $14,608$ edges. As in the case of the yeast data set, the PPI graph of the human data is very sparse, the interacting pairs represent less then 1% from the possible links in the graph.

Both data sets are highly unbalanced, with 1% and 5% positive pairs for yeast data and human data, respectively. There are classification methods that were designed to address the unbalancedness of data [19]. Specifically, for protein interactions, there are some studies [32,20] that investigate how to construct non-interacting protein pairs (negative samples).

3.2   Experimental Setup

The experimental setup considered a part of the data for training and the rest for testing. The training data was used to learn the models and the testing data was used to evaluate how good these models can predict PPIs. We randomly sampled a training set containing 1%, 5%, 10% and 20% protein pairs and their labels as interacting or not from the yeast and human data set. The PPI prediction problem was thus transformed in a binary classification problem The training features were standardized to have mean zero and standard deviation of one. This data sample was used to train the classification model (i.e., learn the weight parameter of the logistic regression). The remaining protein pairs were used for testing the performance. These steps were repeated 10 times and average results are reported (mean $\pm$ standard deviation).

3.3   Evaluation Measure

Area under the receiver operating characteristic curve (AUC) was used as a measure for evaluating the performance. The receiver operator characteristic (ROC) curve plots the true positive rate against the false positive rate for different thresholds. The AUC statistic can be interpreted as the probability that a randomly chosen missing edge (a true positive) is given a higher score by the method than a randomly chosen pair of proteins without an interaction (a true negative).

Table 1. AUC values (mean $\pm$ standard deviation) for the four models: Model 1 represents the Bayesian framework for combining feature and topology information, Model 2 uses only protein information, Model 3 uses only topology information, Mode 4 uses protein features which are enriched by node degrees. The $*$ indicates that the results obtained for Model 1 are significantly better than the results obtained for Model 2. The four upper rows correspond to the yeast data set while the four lower rows correspond to the human data set.

% Train data	Model 1 Features+ Topology	Model 2 Features only	Model 3 Topology only	Model 4 Topology features
1%	0.639 ± 0.014	0.639 ± 0.018	0.577 ± 0.016	0.582 ± 0.022
5%	0.708 ± 0.006	0.697 ± 0.009	0.688 ± 0.010	0.689 ± 0.009
10%	$0.731 \pm 0.005^*$	0.712 ± 0.005	0.720 ± 0.006	0.717 ± 0.007
20%	$0.746 \pm 0.009^*$	0.719 ± 0.006	0.742 ± 0.009	0.737 ± 0.010
1%	$0.863 \pm 0.006^*$	0.851 ± 0.006	0.608 ± 0.014	0.822 ± 0.012
5%	$0.909 \pm 0.002^*$	0.859 ± 0.001	0.793 ± 0.007	0.899 ± 0.003
10%	$0.931 \pm 0.002^*$	0.861 ± 0.001	0.864 ± 0.005	0.931 ± 0.002
20%	$0.952 \pm 0.002^*$	0.862 ± 0.001	0.917 ± 0.003	0.954 ± 0.002

3.4 Performance

Table 1 shows the comparison of the performance of the four models discussed. Model 1 represents the Bayesian framework for combining feature and topology information, Model 2 uses only protein information, Model 3 uses only topology information and Model 4 uses protein features which are enriched with node degrees. The comparison was performed for the yeast data (the four upper rows in Table 1) and human data sets (the four lower rows in Table 1). The protocol described in Section 3.2 was used and the averaged AUC scores with their standard deviations are reported. The statistical significance between Model 1 and Model 2 was assessed by using a Mann-Whitney U-test [8] on the AUC values obtained from the two models for 10 random splits of the data into training and testing. A 5% significance level has been considered. The $*$ indicates that the results obtained for Model 1 are significantly better than the results obtained for Model 2.

The results show that the combination of the two sources of information, protein features and topology, gives a better performance than using only one type of information. In particular Model 1 (Features+Topology) performs significantly better than Model 2 (Features only) in most of the cases. Model 1 and Model 4 have a similar performance for human data, and Model 1 performs better than Model 4 for yeast data. An explanation for this is related to how the protein features were constructed in the two cases; for yeast data the features for a protein pair resulted from summing the feature vectors corresponding to the two proteins, while for human data the protein features are more related to the protein pair than to individual proteins. Model 3 (Topology only) uses only the information related to the topology, in particular the property of hub-proteins to interact with many other proteins. Note that you can have the pair of protein A and protein B in training set and the pair of protein A and protein C in the test set, and in this way the algorithm learns which proteins are hubs (and other topological information) and makes predictions based on topology.

The results vary also as a function of the size of the training data. For a small training set the network is not well defined, and we can see that in this case the improvement is smaller, but, as we increase the training set, meaning that the knowledge about the network topology increases, the performance obtained by adding the topology information improves more.

4   Conclusion

We introduced a framework for predicting PPI by considering the network structure information. This is a Bayesian framework consisting of a prior distribution over the network topology and likelihood terms for observations about links in the network. In the Bayesian framework in general, and in our case when trying to add topological information, the computational complexity is an issue. In the framework presented here, we managed to find some simplifying assumptions which reduce the computational complexity and at the same time yield a good performance.

References

1. Ben-Hur, A., Noble, W.S.: Kernel methods for predicting protein–protein interactions. Bioinformatics 21(1), 38–46 (2005)
2. Chen, X.W., Liu, M.: Prediction of protein-protein interactions using random decision forest framework. Bioinformatics 21(24), 4394–4400 (2005)
3. Chung, F., Lu, L.: Connected components in random graphs with given expected degree sequences. Annals of Combinatorics 6(2), 125–145 (2002)
4. Friedel, C., Zimmer, R.: Inferring topology from clustering coefficients in protein-protein interaction networks. BMC Bioinformatics 7, 519 (2006)
5. Geurts, P., Touleimat, N., Dutreix, M., d'Alché-Buc, F.: Inferring biological networks with output kernel trees. BMC Bioinformatics (PMSB 2006 Special Issue) 8(suppl. 2), S4 (2007)
6. Geurts, P., Wehenkel, L., d'Alché-Buc, F.: Gradient boosting for kernelized output spaces. In: Proceedings of the 24th International Conference on Machine Learning. ACM International Conference Proceeding Series, vol. 227, pp. 289–296. ACM (2007)
7. Geurts, P., Wehenkel, L., d'Alché Buc, F.: Kernelizing the output of tree-based methods. In: Proceedings of the 23th International Conference on Machine Learning, pp. 345–352 (2006)
8. Hollander, M., Wolfe, D.: Nonparametric Statistical Methods. John Wiley & Sons (1999)

9. Jansen, R., Yu, H., et al.: A Bayesian networks approach for predicting protein-protein interactions from genomic data. Science 302(5644), 449–453 (2003)
10. Jeong, H., Mason, S.P., Barabási, A.-L., Oltvai, Z.N.: Lethality and centrality in protein networks. Nature 411(6833), 41–42 (2001)
11. Kashima, H., Yamanishi, Y., Kato, T., Sugiyama, M., Tsuda, K.: Simultaneous inference of biological networks of multiple species from genome-wide data and evolutionary information. Bioinformatics 25(22), 2962–2968 (2009)
12. Kuchaiev, O., Rasajski, M., Higham, D.J., Przulj, N.: Geometric de-noising of protein-protein interaction networks. PLOS Computational Biology 5(8) (2009)
13. Li, Z.C., Lai, Y.H., et al.: Identifying functions of protein complexes based on topology similarity with random forest. Mol. Biosyst. (10), 514–525 (2014)
14. Lin, N., Wu, B., Jansen, R., Gerstein, M., Zhao, H.: Information assessment on predicting protein-protein interactions. BMC Bioinformatics 5, 154 (2004)
15. Maslov, S., Sneppen, K.: Specificity and stability in topology of protein networks. Science 296, 910–913 (2002)
16. Memisevic, V., Milenkovic, T., Przulj, N.: Complementarity of network and sequence information in homologous proteins. Journal of Integrative Bioinformatics 7(3), 135 (2010)
17. Milenkovic, T., Przulj, N.: Uncovering biological network function via graphlet degree signatures. Cancer Informatics 6, 257–273 (2008)
18. Mohamed, T.P., Carbonell, J.G., Ganapathiraju, M.K.: Active learning for human protein-protein interaction prediction. BMC Bioinformatics 11(suppl. 1), S57 (2010)
19. Muntean, M., Valean, H., Ileana, I., Rotar, C.: Improving classification with support vector machine for unbalanced data. In: Proceedings of 2010 IEEE International Conference on Automation, Quality and Testing, Robotics, THETA, 17th edn., pp. 234–239 (2010)
20. Park, Y., Marcotte, E.M.: Revisiting the negative example sampling problem for predicting protein-protein interactions. Bioinformatics 27(21), 3024–3028 (2011)
21. Przulj, N., Corneil, D., Jurisica, I.: Modeling interactome: scale-free or geometric? Bioinformatics 20(18), 3508–3515 (2004)
22. Qi, Y., Klein-Seetharaman, J., Bar-Joseph, Z.: Random forest similarity for protein-protein interaction prediction from multiple sources. In: Altman, R.B., Jung, T.A., Klein, T.E., Dunker, A.K., Hunter, L. (eds.) Pacific Symposium on Biocomputing. World Scientific (2005)
23. Qi, Y., Klein-Seetharaman, J., Bar-Joseph, Z.: A mixture of feature experts approach for protein-protein interaction prediction. BMC Bioinformatics 8(suppl. 10), S6 (2007)
24. Qi, Y., Tastan, O., Carbonell, J.G., Klein-Seetharaman, J., Weston, J.: Semi-supervised multi-task learning for predicting interactions between hiv-1 and human proteins. Bioinformatics 26(18), i645–i652 (2010)
25. Sarajlic, A., Janjic, V., Stojkovic, N., Radak, D., Przulj, N.: Network topology reveals key cardiovascular disease genes. PLoS One 8(8), e71537 (2013)
26. Shi, M.G., Xia, J.F., Li, X.L., Huang, D.S.: Predicting protein-protein interactions from sequence using correlation coefficient and high-quality interaction dataset. Amino Acids 38(3), 891–899 (2010)
27. Sprinzak, E., Altuvia, Y., Margalit, H.: Characterization and prediction of protein-protein interactions within and between complexes. PNAS 103(40), 14718–14723 (2006)
28. Tanaka, R., Yi, T.M., Doyle, J.: Some protein interaction data do not exhibit power law statistics. FEBS Letters 579, 5140–5144 (2005)

29. Tastan, O., Qi, Y., Carbonell, J.G., Klein-Seetharaman, J.: Prediction of interactions between hiv-1 and human proteins by information integration. In: Proceedings of the Pacific Symposium on Biocomputing, vol. 14, pp. 516–527 (2009)
30. von Mering, C., Krause, R., Snel, B., Cornell, M., Oliver, S.G., Fields, S., Bork, P.: Comparative assessment of large-scale data sets of protein-protein interactions. Nature 417(6887), 399–403 (2002)
31. Yamanishi, Y., Vert, J.-P., Kanehisa, M.: Protein network inference from multiple genomic data: a supervised approach. Bioinformatics 20(1), 363–370 (2004)
32. Yu, J., Guo, M., Needham, C.J., Huang, Y., Cai, L., Westhead, D.: Simple sequence-based kernels do not predict protein-protein interactions. Bioinformatics 26(20), 2610–2614 (2010)
33. Zhang, L.V., Wong, S., King, O., Roth, F.: Predicting co-complexed protein pairs using genomic and proteomic data integration. BMC Bioinformatics 5, 38 (2004)

Biases of Drug–Target Interaction Network Data

Twan van Laarhoven and Elena Marchiori

Institute for Computing and Information Sciences, Radboud University Nijmegen,
The Netherlands
{tvanlaarhoven,elenam}@cs.ru.nl

Abstract. Network based prediction of interaction between drug compounds and target proteins is a core step in the drug discovery process. The availability of drug–target interaction data has boosted the development of machine learning methods for the *in silico* prediction of drug–target interactions. In this paper we focus on the crucial issue of data bias.

We show that four popular datasets contain a bias because of the way they have been constructed: all drug compounds and target proteins have at least one interaction and some of them have only a single interaction. We show that this bias can be exploited by prediction methods to achieve an optimistic generalization performance as estimated by cross-validation procedures, in particular leave-one-out cross validation. We discuss possible ways to mitigate the effect of this bias, in particular by adapting the validation procedure. In general, results indicate that the data bias should be taken into account when assessing the generalization performance of machine learning methods for the *in silico* prediction of drug–target interactions.

The datasets and source code for this article are available at
http://cs.ru.nl/~tvanlaarhoven/bias2014/

1 Introduction

An important problem in pharmacology is to find interactions between drug compounds and target proteins in order to understand and study their effects. The *in silico* prediction of such interactions is crucial for improving the efficiency of the laborious and costly experimental determination of drug–target interaction, see e.g. [5].

Drug-target interaction data are publicly available for various classes of pharmaceutically useful target proteins including enzymes, ion channels, GPCRs (G Protein-Coupled Receptors) and nuclear receptors [13]. Various databases have been built and maintained, such as KEGG BRITE [16], DrugBank [29], GLIDA [23], SuperTarget and Matador [12], BRENDA [26], and ChEMBL [24], containing drug–target interaction and other related sources of information, like chemical and genomic data.

The availability of these data stimulated the development of machine learning methods for the *in silico* prediction of drug-target interactions [8]. The current state-of-the-art for the *in silico* prediction of drug–target interaction involves

M. Comin et al. (Eds.): PRIB 2014, LNBI 8626, pp. 23–33, 2014.
© Springer International Publishing Switzerland 2014

methods that employ similarity measures for drug compounds and for target proteins in the form of kernel functions, e.g., Bleakley et al. [2], Chen et al. [4], Gönen [11], vanLaarhoven et al. [21], Mei et al. [22], Wassermann et al. [28], Yamanishi et al. [30, 31].

One can distinguish between prediction for 'known' drug compounds or targets, for which at least one interaction is present in the training set; and prediction for 'unseen' drug compounds or targets, for which no interaction is available in the training set. This results in four possible settings for predicting drug-target interaction, depending on whether the drug compounds and/or targets are known or unseen [30].

In our recent work on predicting drug-target interactions [20] we discovered that a positive bias was implicity introduced in a published method. This motivated the two main research questions we will address in this paper.

1. How does data bias affect the results of procedures used to estimate the generalization performance of a method?
2. Can we quantify and avoid such bias?

Cross-validation (CV) [19] is typically used to assess the generalization performance of methods in the above mentioned settings. The dataset is repeatedly partitioned into two disjoint parts, a training set and a hold-out set. For each partition, the training set is used to construct the predictor and the hold-out set is used for testing. Popular variants are 10-fold CV, where the data is partitioned into ten folds, and each fold is used once as the hold-out set, and leave-one-out cross-validation (LOOCV), where each example constitutes one hold-out set. In the context of drug-target interaction various cross-validation settings can be defined, depending on what is considered an example (e.g. a single drug-target pair or all interactions with a single drug compound) and on the selected CV procedure.

We consider the four popular drug-target interaction datasets in humans involving enzymes, ion channels, G-protein-coupled receptors (GPCRs) and nuclear receptors from Yamanishi et al. [30]. These data have been used as benchmark datasets in recent works, e.g. Bleakley et al. [2], Chen et al. [4], Gönen [11], vanLaarhoven et al. [21, 20], Mei et al. [22], Wassermann et al. [28].

In this paper we show experimentally that these datasets contain a bias which may lead to optimistic CV generalization results. Furthermore, the extent to which this bias affects the results can differ for different methods. As a result, it is unclear whether a method with better CV results on these datasets will also have better performance in real applications.

Specifically, these datasets have been constructed in such a way that each drug compound and target protein has at least one interaction. Furthermore, some drug compound and/or targets have only a single interaction.

We show how this bias can be incorporated into a baseline prediction method in such a way that it significantly increases the LOOCV generalization performance. We investigate how this bias can be reduced and quantified. We show experimentally that 5- or 10-fold CV reduces (but does not eliminate) the bias. Furthermore, the presence of this bias can be quantified by separating

the performance metrics for drug compounds and targets with just one interaction from that for other drug–target interaction pairs. This provides an alternative procedure to assess the generalization performance of a method by
highlighting the effect of the data bias.

In general, our results provide a contribution towards the understanding of CV
procedures in the presence of data bias in the context of drug-target interaction
networks.

1.1   Related Work

Dataset bias has been investigated in different domains, e.g. in ligand based
virtual screening [1], where local clustering and global spread of the considered
benchmark datasets were identified influencing validation results, and in object
recognition [27], where current state of recognition datasets have been compara-
tively analyzed and evaluated based on criteria including relative data bias and
cross-dataset generalization. To the best of our knowledge, this is the first time
that drug–target interaction network data bias is analyzed.

The dangers of CV have been studied by the machine learning community
in various contexts. For instance, in Isaksson et al. [14] CV and bootstrapping in small sample classification are investigated. A fundamental problem is
that the uncertainty in a point estimate obtained with these procedures is unknown and may be quite large. The authors therefore suggest that the final
classification performance should be reported in the form of a Bayesian confidence interval or using some other method that yields conservative measures of
the uncertainty. Furthermore, in Rao et al. [25] it was empirically shown that
when the number of algorithms is large, LOOCV is not an effective estimate of
generalization performance for the algorithm that has the best cross-validation
performance. The authors showed that this behavior worsens as the sample size
decreases, and as the dimensionality and number of algorithms increase. The
phenomenon of under-estimating cross validation error was also demonstrated
on some benchmark data sets, and was shown to be worse for datasets with
higher dimensionality.

2   Materials

In Yamanishi et al. [30] datasets were introduced for the drug–target prediction problem. These datasets are based on four different domains: enzymes, ion
channels, GPCRs and nuclear receptors. The datasets are constructed in such a
way that only the proteins that have an interacting drug are included, and for
each domain only the drugs that interact with at least one protein are included.
It turns out that this property introduces problems for validation.

In Table 1 we give an overview of the four datasets as they are used in recent
publications. As can be seen in the last column, a large fraction of the drug
compounds and target have just one interaction in the dataset. Or equivalently,
there are many interactions which are the only interaction for a drug–target. We
call such interacting pairs *unique*.

The interactions in a dataset can be encoded in a matrix y_{dt}, such that $y_{dt} = 1$ if drug compound d interacts with target protein t, and $y_{dt} = 0$ otherwise. Besides this interaction information, there is also other information available on the drugs and targets themselves. Usually this is encoded in kernel matrices that give a similarity score between two drugs or two targets.

Table 1. The number of drug compounds, the number of target proteins, the number of interactions and the number of unique interaction pairs (interactions which are the only one for a drug or target) in the drug–target datasets from Yamanishi et al. [30]

Dataset	Drugs	Targets	Interactions	Unique
Enzyme	445	664	2926	451 (15%)
Ion Channel	210	204	1476	103 (7%)
GPCR	223	95	635	132 (21%)
Nuclear Receptor	54	26	90	44 (49%)

3 Methods

3.1 Validation Procedures

There are two main ways in which these datasets of interactions can be used by machine learning methods:

1. To train a model to predict with which targets a previously unseen drug will interact. We call this the 'unseen drug' setting.
2. To find new putative interactions between drugs and targets already in the dataset. We call this the 'pairs' setting.

An overview of the prediction setting and type of CV used in state-of-the-art methods applied to these datasets are shown in Table 2. In this work we focus primarily on the 'pairs' setting, which is used by most of the methods listed in the table.

Usually methods are compared by looking at the ranking of interactions they produce in a cross-validation setting. That is, each drug–target pair is assigned a score by each method, where only other interacting pairs are shown to the method. Then the pairs are ranked based on these scores and the quality of the ranking is compared using AUC, AUPR or other summary statistics. Specifically, the ROC curve of true positives as a function of false positives is computed, and the area under the ROC curve (AUC) is considered as quality measure, see for instance [10]. Furthermore, the precision–recall curve is computed, that is, the plot of the ratio of true positives among all positive predictions for each given recall rate. The area under this curve (AUPR) is a more informative quality measure than the AUC, as it punishes much more the existence of false positive examples found among the top ranked prediction scores [6].

Table 2. A list of papers that used the interaction data in Table 1, showing the type of prediction setting ('unseen drug' or 'pairs') and type of CV procedure used

	Unseen drug	Pairs	CV procedure
Yamanishi et al. [30]	✓	✓	10-fold CV
Bleakley et al. [2]	✓	✓	LOOCV, 10-fold CV
vanLaarhoven et al. [21]	-	✓	LOOCV, 10-fold CV
Chen et al. [4]	-	✓	LOOCV
Gönen [11]	✓	-	5-fold CV
Mei et al. [22]	-	✓	LOOCV, 10-fold CV
vanLaarhoven et al. [20]	✓	✓	LOOCV, 5-fold CV

3.2 Biases

Suppose that a method is tested using LOOCV. Then if a unique interaction (d, t) is left out, the method will see a row (or column) of zeros in the matrix. But we know that the dataset does not have such rows or columns, since each drug and target has at least one interaction. We can therefore know with certainty that this pair interacts. This process is illustrated in Fig. 1.

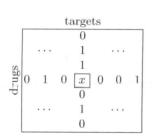 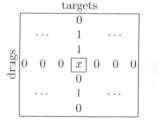

Fig. 1. In the LOOCV procedure, the task is to predict a single unknown drug–target interaction, assuming all other interactions are known. This is indicated by x in the matrix of drug–target interactions. Because of the construction of the dataset, we can know with certainty that in the second matrix $x = 1$, otherwise this drug compound would not be included in the dataset.

Consider a simple baseline method, that ranks drug–target pairs by the number of adjacent pairs that are known to interact, where two drug–target pairs are adjacent if they share a drug or a target. This number of adjacent interacting pairs for the pair (d, t) is

$$a_{dt} = a_{dt}^{\mathrm{drug}} + a_{dt}^{\mathrm{targ}}, \quad \text{where} \quad a_{dt}^{\mathrm{drug}} = \sum_{d' \neq d} y_{d't}, \quad a_{dt}^{\mathrm{targ}} = \sum_{t' \neq t} y_{dt'}.$$

At first glance we would expect drugs or targets that already have many known interactions to be more promiscuous, and therefore also more likely to interact with other drugs and targets. But as explained in the previous paragraph that is not the case when LOOCV is used.

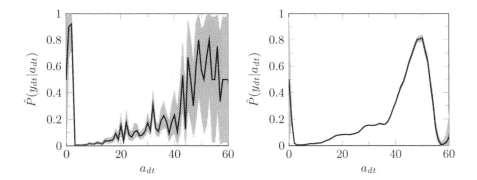

Fig. 2. Probability of a drug–target pair interacting given the number of adjacent interactions. The first plot shows this probability for LOOCV in the GPCR dataset, the second plot for 10-fold cross validation. The shaded area indicate a 95% confidence interval based on a uniform prior.

To test this effect, in Fig. 2 we have plotted the fraction of pairs that interact against a_{dt}. This is an empirical estimate $\hat{P}(y_{dt}|a_{dt})$ of the probability that d and t interact given the number of adjacent pairs for (d, t). Overall there is indeed a trend for larger a_{dt} to correspond to a higher probability of interacting. But for very low a_{dt} we see the bias in action: the probability of such pairs interacting is very high, since many of them are unique interactions.

A method can exploit this knowledge as follows. Consider the biased variant of the baseline method, which is the same as the baseline, except that it ranks the pairs with no observed adjacent pairs sharing a drug or with no pairs sharing a target before all other drug–target pairs. More precisely, instead of ranking pairs by a_{dt}, they are ranked by

$$a_{dt}^{\text{unique} \mapsto \infty} = \begin{cases} \infty & \text{if } a_{dt}^{\text{drug}} = 0 \text{ or } a_{dt}^{\text{targ}} = 0 \\ a_{dt} & \text{otherwise.} \end{cases}$$

In Table 3 we compare the LOOCV performance of this biased method to the unbiased baseline.

To estimate the statistical significance of the AUC results we used the method described in DeLong et al. [7]. To determine significance of the AUPR results we used bootstrapping.

The difference between the unbiased and the biased methods is purely due to the unique interactions. In Table 3 we also show the AUC and AUPR split up for just the unique and non-unique interactions. With the unbiased baseline method, the AUC for unique interactions is barely above random chance level, while the biased baseline method achieves a perfect AUC. The overall AUC is

Table 3. Performance of the unbiased baseline method and the biased variant when tested with LOOCV. The best results for each dataset are indicated in bold.

Dataset	Method	AUC			AUPR		
		overall	unique	other	overall	unique	other
Enzyme	Baseline	0.880	0.668	0.919	0.101	0.006	0.102
	Biased	**0.931**	1.000	0.919	**0.301**	1.000	0.102
Ion Channel	Baseline	0.850	0.528	0.874	0.244	0.003	0.254
	Biased	**0.883**	1.000	0.874	**0.355**	1.000	0.254
GPCR	Baseline	0.796	0.542	0.863	0.157	0.009	0.168
	Biased	**0.891**	1.000	0.863	**0.420**	1.000	0.168
Nuclear Receptor	Baseline	0.703	0.511	0.887	0.152	0.044	0.143
	Biased	**0.942**	1.000	0.887	**0.682**	1.000	0.143

a weighted average of the AUCs for unique and non-unique interactions, where the weight corresponds to the fraction of unique interactions. For example, for the GPCR dataset, $79\% \cdot 0.863 + 21\% \cdot 1.000 = 0.891$. Such a relation does not hold for AUPR scores, but the overall picture is similar.

4   Avoiding the Bias

It seems that the biased results stem from the use of LOOCV. And so one would hope to avoid this problem by using 10 fold CV instead. As the right part of Fig. 2 shows, this indeed reduces the bias, but it does not completely eliminate it.

We have repeated the experiment from the previous section with 10-fold CV instead of LOOCV. This is the setting used by, for instance Yamanishi et al. [30]. As seen in the Table 4, exploiting the data bias still improves the AUC and AUPR scores for unique interactions, but this comes at the cost of the performance for non-unique interactions. In general, with k-fold cross-validation on a dataset with n drugs/targets, for each unique interacting pair, there are on the order of n/k non-interacting pairs that will be excluded in the same fold. These pairs will appear similar to the unique interaction ones. As the dataset becomes larger, there will be more such pairs.

However, it is still possible to beat the baseline method by making a trade-off between the increased performance on unique interactions and decreased performance on other interactions. For example, one can introduce the 'slight bias' method that ranks pairs which appear to be unique as if they have k adjacent interactions. So it ranks pairs by $a_{dt}^{\text{unique}\mapsto k}$ for some $k < \infty$. By tuning this parameter k we can tune the trade-off. In our experiments we chose k with cross validation. As shown in Table 4, this method achieves best AUC and AUPR on all but the smallest dataset; and in all cases shows a significant improvement over the baseline method.

Table 4. Performance of the unbiased baseline method and the biased variants when tested with 10 fold CV. The best results for each dataset are indicated in bold, results in italic do not differ significantly from the best (at $\alpha = 0.05$).

Dataset	Method	AUC			AUPR		
		overall	unique	other	overall	unique	other
Enzyme	Baseline	0.879	0.669	0.917	0.098	0.006	0.099
	Slight bias	**0.900**	0.818	0.915	**0.101**	0.012	0.097
	Biased	0.862	0.982	0.840	0.056	0.135	0.027
Ion Channel	Baseline	0.849	0.530	0.873	0.246	0.003	0.254
	Slight bias	**0.859**	0.695	0.871	**0.248**	0.005	0.252
	Biased	0.836	0.987	0.824	0.123	0.128	0.098
GPCR	Baseline	0.795	0.543	0.859	0.154	0.009	0.163
	Slight bias	**0.841**	0.801	0.853	**0.168**	0.025	0.155
	Biased	0.827	0.975	0.788	0.116	0.180	0.057
Nuclear Receptor	Baseline	0.697	0.533	0.885	0.154	0.047	0.155
	Slight bias	0.857	0.884	0.846	0.247	0.177	0.124
	Biased	**0.878**	0.967	0.781	**0.351**	0.473	0.070

So far we have considered the bias in the pairs setting. Results suggest that perhaps this validation setting should not be used. An alternative is the unseen drug setting, where one or more rows are left out in their entirety from the drug–target interaction matrix. This means that it becomes impossible to see if a pair is unique for a certain drug. But there are still interactions that are unique for a target. As shown in Table 5, this bias can still be exploited for improving CV performance, even when using 5- or 10-fold cross-validation.

Another option is to separate the unique interactions from the non-unique interactions when doing validation. As shown in our experiments, the non-unique interactions are not sensitive to the same bias. A good solution would be to only consider the AUC and AUPR scores for the non-unique interactions when comparing different methods. This still introduces a bias of a different kind, however, since some drug compounds and targets will be unnecessarily excluded.

A different way to validate a method is to seek confirmation of the predictions in other datasets. This is done by for instance Yamanishi et al. [31], van-Laarhoven et al. [21], Gönen [11], where the 10 highest rank predictions are looked up in the literature, and in newer versions of the KEGG BRITE, Drug-Bank Chembl, SuperTarget and Matador databases. A problem with such validation is that it is hard to quantify the performance, because only a few interactions are verified, and because these databases are extended between the publication of different papers.

Perhaps the most principled way of avoiding biases in validation is to act on the data and construct more realistic datasets. For this problem, that means that the dataset should also include compounds that interact with none of the targets,

Table 5. Performance of the baseline method and biased variants in the unseen drug setting, when validated with 5-fold CV. The best results for each dataset are indicated in bold, results in italic do not differ significantly from the best (at $\alpha = 0.05$).

Dataset	Method	AUC			AUPR		
		overall	unique	other	overall	unique	other
Enzyme	Baseline	0.723	0.320	0.802	0.040	0.003	0.039
	Slight bias	**0.772**	0.637	0.814	**0.041**	0.003	0.040
	Biased	0.747	0.868	0.743	0.023	0.018	0.016
Ion Channel	Baseline	0.699	0.602	0.710	*0.079*	0.010	0.075
	Slight bias	**0.701**	0.677	0.707	**0.080**	0.010	0.075
	Biased	*0.698*	0.797	0.694	0.064	0.017	0.059
GPCR	Baseline	0.766	0.562	0.819	0.094	0.012	0.088
	Slight bias	**0.782**	0.664	0.813	**0.095**	0.013	0.087
	Biased	0.750	0.747	0.747	0.062	0.025	0.047
Nuclear Receptor	Baseline	0.616	0.585	0.650	*0.140*	0.067	0.109
	Slight bias	*0.647*	0.633	0.653	**0.144**	0.070	0.109
	Biased	**0.670**	0.699	0.626	*0.126*	0.084	0.059

or targets for which there is no known interacting compound. The question then becomes which other drug compounds and proteins to include in the dataset. This possibility remains to be explored.

5   Conclusions

We have shown that popular benchmark data for the drug–target interaction problem are biased because they include only drug compounds and target proteins with at least one interaction. This bias can be quantified by looking at the CV performance on these unique interactions separately from non-unique interactions. The bias is the largest with leave-one-out cross-validation in the pairs setting. But even with 5- or 10-fold cross-validation and in the unseen drugs setting there is still a significant bias. Our analysis indicates that results of CV procedures to assess the predictive performance of methods for drug–target interaction networks should be interpreted with care because they could be possibly positively affected by bias contained in the considered datasets.

The baseline method discussed in this paper does not use the similarity information of drug compounds or target proteins at all. Hence, the performance is far below the state of the art. However, the effects of the bias carry over to other methods. For any ranking method r_{dt} we can define a variant $r_{dt}^{\text{unique} \mapsto k}$ that exploits the dataset bias and thereby boosts the performance on unique interacting pairs.

We have not performed an empirical study of the prevalence of biases in published methods. Of course none of the methods in Table 2 exploit the bias in quite such a blatant way as our 'biased baseline' method. Still, there could be methods that inadvertently take more advantage of the bias than others, for example in the choice of parameter values or in the way they handle specific types of drug–target pairs.

In this work we have focused on a single group of datasets, with a specific type of interaction, drug–target interaction. It remains to be investigated whether other datasets for the drug–target interaction prediction problem and datasets for other similar problems have the same bias. It would also be interesting to consider other interaction datasets, such as the drug–target, enzyme–motabolite and protein–ligand datasets from [17, 3, 9, 15, 18].

References

1. Baumann, K., Rohrer, S.: Exploring benchmark dataset bias in ligand based virtual screening. Chemistry Central Journal 2(suppl. 1), P1 (2008)
2. Bleakley, K., Yamanishi, Y.: Supervised prediction of drug-target interactions using bipartite local models. Bioinformatics 25(18), 2397–2403 (2009)
3. Campillos, M., Kuhn, M., Gavin, A.-C., Jensen, L.J., Bork, P.: Drug target identification using side-effect similarity. Science 321(5886), 263–266 (2008)
4. Chen, X., Liu, M.-X., Yan, G.-Y.: Drug-target interaction prediction by random walk on the heterogeneous network. Mol. Biosyst. 8(7), 1970–1978 (2012)
5. Csermely, P., Korcsmáros, T., Kiss, H.J., London, G., Nussinov, R.: Structure and dynamics of molecular networks: A novel paradigm of drug discovery: A comprehensive review. Pharmacology & Therapeutics 138(3), 333–408 (2013)
6. Davis, J., Goadrich, M.: The relationship between Precision-Recall and ROC curves. In: ICML 2006: Proceedings of the 23rd International Conference on Machine Learning, pp. 233–240. ACM (2006)
7. DeLong, E.R., DeLong, D.M., Clarke-Pearson, D.L.: Comparing the Areas under Two or More Correlated Receiver Operating Characteristic Curves: A Nonparametric Approach. Biometrics 44(3), 837–845 (1988)
8. Ding, H., Takigawa, I., Mamitsuka, H., Zhu, S.: Similarity-based machine learning methods for predicting drug–target interactions: a brief review. Briefings in Bioinformatics (2013)
9. Faulon, J.-L., Misra, M., Martin, S., Sale, K., Sapra, R.: Genome scale enzyme–metabolite and drug–target interaction predictions using the signature molecular descriptor. Bioinformatics 24(2), 225–233 (2008)
10. Fawcett, T.: An introduction to ROC analysis. Pattern Recognition Letters 27(8), 861–874 (2006)
11. Gönen, M.: Predicting drug-target interactions from chemical and genomic kernels using Bayesian matrix factorization. Bioinformatics 28(18), 2304–2310 (2012)
12. Günther, S., Kuhn, M., Dunkel, M., Campillos, M., Senger, C., Petsalaki, E., Ahmed, J., Urdiales, E.G.G., Gewiess, A., Jensen, L.J.J., Schneider, R., Skoblo, R., Russell, R.B., Bourne, P.E., Bork, P., Preissner, R.: SuperTarget and Matador: resources for exploring drug-target relationships. Nucleic Acids Res. 36(Database issue), D919–D922 (2008)
13. Hopkins, A.L., Groom, C.R.: The druggable genome. Nature reviews. Drug Discovery 1(9), 727–730 (2002)
14. Isaksson, A., Wallman, M., Göransson, H., Gustafsson, M.G.: Cross-validation and bootstrapping are unreliable in small sample classification. Pattern Recognition Letters 29(14), 1960–1965 (2008)

15. Jacob, L., Hoffmann, B., Stoven, B., Vert, J.-P.: Virtual screening of GPCRs: an in silico chemogenomics approach. BMC Bioinformatics 9, 363 (2008)
16. Kanehisa, M., Goto, S., Hattori, M., Aoki-Kinoshita, K.F., Itoh, M., Kawashima, S., Katayama, T., Araki, M., Hirakawa, M.: From genomics to chemical genomics: new developments in KEGG. Nucleic Acids Res. 34(Database issue), D354–D357 (2006)
17. Keiser, M.J., Roth, B.L., Armbruster, B.N., Ernsberger, P., Irwin, J.J., Shoichet, B.K.: Relating protein pharmacology by ligand chemistry. Nat. Biotechnol. 25(2), 197–206 (2007)
18. Keiser, M.J., Setola, V., Irwin, J.J., Laggner, C., Abbas, A.I., Hufeisen, S.J., Jensen, N.H., Kuijer, M.B., Matos, R.C., Tran, T.B., Whaley, R., Glennon, R.A., Hert, J., Thomas, K.L., Edwards, D.D., Shoichet, B.K., Roth, B.L.: Predicting new molecular targets for known drugs. Nature 462(7270), 175–181 (2009)
19. Kohavi, R.: A study of cross-validation and bootstrap for accuracy estimation and model selection. In: Proceedings of the 14th International Joint Conference on Artificial Intelligence, IJCAI 1995, vol. 2, pp. 1137–1143. Morgan Kaufmann Publishers Inc., Montreal (1995)
20. van Laarhoven, T., Marchiori, E.: Predicting Drug-Target Interactions for New Drug Compounds Using a Weighted Nearest Neighbor Profile. PLoS One 8(6), e66952 (2013)
21. van Laarhoven, T., Nabuurs, S.B., Marchiori, E.: Gaussian interaction profile kernels for predicting drug–target interaction. Bioinformatics 27(21), 3036–3043 (2011)
22. Mei, J.-P., Kwoh, C.-K., Yang, P., Li, X., Zheng, J.: Drug-target interaction prediction by learning from local information and neighbors. Bioinformatics 29(2), 238–245 (2013)
23. Okuno, Y., Tamon, A., Yabuuchi, H., Niijima, S., Minowa, Y., Tonomura, K., Kunimoto, R., Feng, C.: GLIDA: GPCR ligand database for chemical genomics drug discovery database and tools update. Nucleic Acids Research 36(suppl. 1), D907–D912 (2008)
24. Overington, J.: ChEMBL. An interview with John Overington, team leader, chemogenomics at the European Bioinformatics Institute Outstation of the European Molecular Biology Laboratory (EMBL-EBI). Interview by Wendy A. Warr. Journal of Computer-Aided Molecular Design 23(4), 195–198 (2009)
25. Rao, R.B., Fung, G.: On the Dangers of Cross-Validation. An Experimental Evaluation. In: SDM, pp. 588–596. SIAM (2008)
26. Schomburg, I., Chang, A., Ebeling, C., Gremse, M., Heldt, C., Huhn, G., Schomburg, D.: BRENDA, the enzyme database: updates and major new developments. Nucleic Acids Res. 32(suppl. 1), D431–D433 (2004)
27. Torralba, A., Efros, A.A.: Unbiased look at dataset bias. In: Proceedings of the 2011 IEEE Conference on Computer Vision and Pattern Recognition, CVPR 2011, pp. 1521–1528. IEEE Computer Society, Washington, DC (2011)
28. Wassermann, A.M., Geppert, H., Bajorath, J.: Ligand prediction for orphan targets using support vector machines and various target-ligand kernels is dominated by nearest neighbor effects. J. Chem. Inf. Model 49, 2155–2167 (2009)
29. Wishart, D.S., Knox, C., Guo, A.C.C., Cheng, D., Shrivastava, S., Tzur, D., Gautam, B., Hassanali, M.: DrugBank: a knowledgebase for drugs, drug actions and drug targets. Nucleic Acids Res. 36(Database issue), D901–D906 (2008)
30. Yamanishi, Y., Araki, M., Gutteridge, A., Honda, W., Kanehisa, M.: Prediction of drug-target interaction networks from the integration of chemical and genomic spaces. Bioinformatics 24, i232–i240 (2008)
31. Yamanishi, Y., Kotera, M., Kanehisa, M., Goto, S.: Drug-target interaction prediction from chemical, genomic and pharmacological data in an integrated framework. Bioinformatics 26(12), i246–i254 (2010)

Logol: Expressive Pattern Matching in Sequences. Application to Ribosomal Frameshift Modeling

Catherine Belleannée, Olivier Sallou, and Jacques Nicolas

Irisa/Inria/Université de Rennes1, Campus de Beaulieu,
35042 Rennes, France
{Catherine.Belleannee,Olivier.Sallou,
Jacques.Nicolas}@irisa.fr

Abstract. Most of the current practice of pattern matching tools is oriented towards finding efficient ways to compare sequences. This is useful but insufficient: as the knowledge and understanding of some functional or structural aspects of living systems improve, analysts in molecular biology progressively shift from mere classification tasks to modeling tasks. People need to be able to express global sequence architectures and check various hypotheses on the way their sequences are structured. It appears necessary to offer generic tools for this task, allowing to build more expressive models of biological sequence families, on the basis of their content and structure.

This article introduces Logol, a new application designed to achieve pattern matching in possibly large sequences with customized biological patterns. Logol consists in both a language for describing patterns, and the associated parser for effective pattern search in sequences (RNA, DNA or protein) with such patterns. The Logol language, based on an high level grammatical formalism, allows to express flexible patterns (with mispairings and indels) composed of both sequential elements (such as motifs) and structural elements (such as repeats or pseudoknots). Its expressive power is presented through an application using the main components of the language : the identification of -1 programmed ribosomal frameshifting (PRF) events in messenger RNA sequences.

Logol allows the design of sophisticated patterns, and their search in large nucleic or amino acid sequences. It is available on the GenOuest bioinformatics platform at *http://logol.genouest.org*. The core application is a command-line application, available for different operating systems. The Logol suite also includes interfaces, e.g. an interface for graphically drawing the pattern.

1 Background

During last decade, a number of pattern matching tools have been proposed, and some of them are used extensively and helpful. Depending on systems under study, modeling needs may vary from looking for all exact occurrences of a given string in a protein bank, to looking for approximated occurrences of a

M. Comin et al. (Eds.): PRIB 2014, LNBI 8626, pp. 34–47, 2014.
© Springer International Publishing Switzerland 2014

given transposon in a full genome, or locating pseudoknots in a RNA sequence. This section proposes a quick overview of the existing software diversity, which shows there exists still room for a new pattern matching tool, both flexible (high "genericity") and with the capacity to represent complex structures (high expressive power).

1.1 General Purpose Pattern Matching

Some tools have been designed for the analysis of several types of sequences (DNA, RNA, proteins) with an generic expressiveness, i.e. without targeting the recognition of a particular motif family. Among these general tools, two tendencies can be observed, efficiency-oriented and expressiveness-oriented software.

One of the most advanced software from the point of view of efficiency is the Vmatch suite (`http://www.vmatch.de`) that offers a large variety of search facilities in very large sequences. It is based on a careful implementation of enhanced suffix trees for the computation of a sequence index that provides a fast access to every substring in that sequence. If the search for a motif contains some rare substrings, this technique is particularly efficient. The software Vmatch is the core search engine used in a number of more specialized tools working on specific sequence structures (e.g. "tandem-repeats" or LTR retrotransposons). Another highly generic tool is Biogrep[11], designed with the objective of *quickly recognizing a large set of simple motifs* (typically more than 100) in biological sequence banks. Biogrep allows queries in the POSIX language, a standard format of extended regular expressions, and can look for patterns in parallel on a set of processors.

The other approach for the analysis of biological sequences is more concerned with modeling the peculiarities of biological objects in the most relevant and expressive way. A major contribution in this respect is the work of D. Searls who laid the foundations for the research in this domain. He was the first to supervise developments allowing users to design biological grammars and to apply them for the large scale analysis of their genomic sequences [19,4,18]. D. Searls has introduced a very practical object in algebraic grammars, the *string variable*, which allows to elegantly express the notion of copy (either direct or reverse). He has implemented the resulting logical formalism, called SVG -for String Variable Grammars-, in the (no longer available) Genlang tool [4]. The direct copy (*e.g.* X...X) allows to search for two occurrences of a same unknown string, using optionally some indication on the string size. The reverse copy (*e.g.* X... ~X) introduces in addition the notion of reverse complement and allows thus the representation of biological palindromes like stem-loops (Stem, Loop, ~Stem) or pseudo-knots (Stem1, Loop1,Stem2, Loop2, ~Stem1, Loop3,~Stem2). Genlang, Stan[15] (developed in our research team), Patscan[5] and Patsearch[16] are all tools belonging to this family. Thanks to string variables and other additional components, these languages offer the possibility to combine easily in a single model informations on the sequence and on the structure of a molecule.

1.2   Dedicated Pattern Matching

It is not possible to provide here an exhaustive review of the profusion of specific tools that have been made available to bioanalysts. Some are specific to a sequence family and others to a particular motif type. A famous one *dedicated to proteins* is ScanProsite[3], where motifs are built upon regular expressions that are searched either by a query in a precomputed database or with the algorithm PS-SCAN[9]. A number of tools are *dedicated to RNA sequences*, in response to the increasing needs of structure exploration in the complex RNA world boosted by the recent importance of non coding RNA studies. For instance, RNAmotif[13], RNAbob[6], Hypasearch[10,20] and Palingol[2] have been designed for the description of patterns as a succession of stems and loops, usually offering the possibility of choosing either a standard Watson-Crick pairing (A-U, G-C) or a pairing including Wobble (A-U, G-C, G-U). A more recent tool in this category, Structator[14], significantly improves the parsing time by making use of an index structure that is suited for the analysis of palindromic structures, the affix arrays. Patterns may also contain some sequence information on words that have to be present in particular places of the stems or the loops. RNAmotif is probablly the most popular in this category.

2   Logol Language

In this landscape, we designed Logol, a new general grammatical pattern matching tool, in order to greatly enhance the range of admissible patterns.

2.1   Basics: A Grammatical Model with Constraints

• **String Variable Grammars:** With the objective of being a general and expressive language allowing a natural expression of composite patterns, the Logol language has been designed on the basis of String Variable Grammars (SVG) introduced by D.Searls [19,4,18]. As already mentioned in the previous section, while it is easy to express motifs and gaps by regular expressions (*e.g.* PROSITE [3]), SVG allow to express structures beyond the capabilities of regular languages such as palindromes (e.g. in stem-loops and pseudo-knots) and repeats (duplicated substrings), which are recorded by string variables. In fact, SVG and Logol grammars lay even beyond the possibilities of context-free grammars (XML-like), in a class that A. Joshi called *"mildly context sensitive languages"* [12]. Starting from the sound basis of SVG grammars, the Logol language proposes several extensions -most notably by adopting a constraint approach- with the goal to allow the expression of realistic biological motifs. The rest of the section introduces its main constituents.

• **First Steps in Logol:** Let us first present a very simple Logol grammar:
```
mod1()==*> SEQ1
mod1()==>"aaa"
```

The first line is the top level instruction (called 'rule'). The rule is identified by the constant '==*> SEQ1' and it triggers the parsing of a sequence for a particular grammatical model (here, mod1). The second line provides the model (i.e. pattern) definition itself, here the string "aaa". It triggers the search for all occurrences of "*aaa*" in a genomic sequence.

The next grammar describes a slightly more interesting pattern made of two distant copies of a same string. The size of the string (in range [5,8]) and the distance between the two copies (in range[1,10]) are bounded but the content of the string itself is kept free:

```
mod2()==*> SEQ1
mod2() ==> X1:#[5,8], .*:{#[1,10]}, X1
```

The model mod2() reads as follows: X1 denotes a string variable; any string made of 5 to 8 letters can be an instance of X1. '.*' denotes a space ('gap'). It is constrained to have a size between 1 and 10 characters. A second occurrence of X1 is waited for after the gap. For example, **acuggc**cc*gacuggc***acuggc** is an instance of this pattern on the input sequence *uucagacuggcccgacuggcacuggccac*, with X1 = *acuggc*.

• **All Matches:** Logol returns all the instances of a model. For instance, when launched on the sequence *cagaaaacgccgaaacuggc* with the model "aaa", it returns the three possible matches: in position 3, 4 and 12.

• **A Powerful Feature: Instance Saving.** The Logol language supports an alternative way to express the pattern mod2. That is:

```
mod2() ==> X1:{#[5,8],_IX1}, .*:{#[1,10]}, ?IX1.
```

In this case, the string corresponding to the occurrence value of X1 is saved (using '_') in a new variable (named here IX1). After a gap of length 1 to 10, the same string IX1 is required again and called back using '?'.

This complicated version of mod2() is only shown for the purpose of introducing the notion of *instance saving* that will be fully used in the next paragraphs. Actually, the various instances of a variable are not necessarily exact copies and this explicit naming process (here _IX1) makes it possible to distinguish one instance from another. Furthermore, such a mechanism allows to save some instance in any part of a model and refer to it elsewhere in the model.

2.2 Constraints

Logol modeling is based on a *constraint approach*. The various constraint types applicable to a model element may be split into two categories, *string constraints* and *structure constraints*. String constraints delimit the start (@), the end (@@), the content (?) and the length (#) of admissible strings. Structure constraints include cost constraints ($ for mismatch count, $$ for indel count) and composition constraints (%). These two categories of constraints are written in two separated sets, as in the following model: mod1() ==> X1:{#[6,7],@[3,11]}:{% "a":50}. Here mod1 looks for a string whose size is in range [6,7], which starts at a position in range [3,11] and contains at least 50% of a. For example, the instances of this pattern in the sequence *ccaaaacgtacgttttttttcccccc* are *aaacgt*, *aaacgta* and *aacgta* (positions start at 0 in a sequence).

• **Non Exact Copies: Mismatch and Indel Cost Constraints ($ and $$)** Genomic sequences evolve through a duplication process prone to errors or mutations. Elementary variations (on one position) between a model and its instances are taken into account through two dedicated cost counters: the counter of *mismatches* (i.e. substitutions) and the counter of *indels*. A mismatch cost constraint is defined by a $[m,n] expression, where m and n are integers. This constraint allows from m to n substitutions. For example, *aaaa*, *acaa* and *aagt* are all instances of the pattern "aaaa":{$[0,2]}.

A mismatch constraint can also take the form of a rate: p$[m,n]. Here, m (resp. n) designates the minimum (resp. maximum) allowed percentage of substitutions. For example, *aaaa*, *acaa* and *aagt* are all instances of the pattern "aaaa":{p$[0,50]}. Indels are defined similarly, by setting the indel cost constraints $$[m,n] and p$$[m,n]. Thus, "aaaa":{$$[0,1]} is accepting, among others, the strings *aaaa* (no indel), *aaa* (one deletion) or *aaaca* (one insertion). Here is a new example to further illustrate the concept of instance saving:

X1:{#[5,8],_I1},.*:{#[1,7]}, ?I1:{_I2}:{$[1,1]}, .*:{#[1,7]}, ?I2:{$[1,1]}

This model allows to look for 3 instances of a same string successively deriving from each other (e.g I1 =aaaaa, I2 = aaaca and I3 = agaca). The second pattern, ?I1: {_I2}:{$[1,1]}, reads as follow: the expected string must be similar to the previous I1 string (aaaaa here), apart from 1 mismatch ($[1,1]). The matched string (aaaca) is saved in I2 ({_I2}) for further use ({?I2}). This individualization of instances allows to adjust fine notions of sequence evolution.

• **Letter Frequencies: Composition Constraints (%)**: Some properties like hydrophobic regions in proteins or GC content in RNA correspond to statistical expectations on a particular segment composition rather than the search of a well-defined element. Logol proposes the expression of *composition constraints* that check the relative frequency of given letters in a sequence. Thus X1:{#[2,43]}:{% "gc":65} describes a segment of length 2 to 43 characters with a GC rate of at least 65%.

2.3 Operators

• **Negation:** Also called *negative content constraint*, negation can be used in order to exclude some values in a motif. It is denoted by the exclamation mark symbol, !. Thus ("aaa"| "ttt"), !"ga":{#[2,2]} refers to a string made of 5 characters, the first three being 3 a or 3 t, and the next two being anything but the word ga.

• **Morphism:** A morphism is a function that applies a transformation to a string by substituting letters or substrings. It can be used in direct (+) or reverse (-) direction. Each user can define its own morphisms, but some are already defined. For instance, "wc" transforms a RNA sequence into its complement

sequence, applying the Watson-Crick pairing (A-U, G-C). Thus, the pattern
+"wc" "acuggc" represents the string "ugaccg" and -"wc" "acuggc" represents the string "gccagu".

The morphism -"wc" produces the *reverse complement* of a string and can be
used to describe biological palindromes such as stem-loops. The next example
provides a pattern for the recognition of stem-loops whose stem length varies
between 5 and 11 and loop size between 1 and 9. Moreover, the Watson-Crick
pairing is not required to be perfect: up to 2 substitutions and 1 indel are allowed.
STEM1:{#[5,11],_IS1}, .*:{#[1,9]}, -"wc" ?IS1 :{$[0,2],$$[0,1]}
In this description, the content of STEM1 (first strand of the stem) is saved in IS1,
(_IS1). The second stem strand is then defined as the exact reverse complement
of the previous content (that is -"wc" ?IS1), except for 2 mismatch and 1 indel.

• **Repeats:** Tandem repeats are frequent genomic structures made of directly
adjacent copies of a same entity that may contain only a few letters (microsatel-
lites) or be longer and reach size over 100 nucleic acids (minisatellites). In Logol,
such structures can be handled by applying a special constructor, repeat, which
manages the characteristics of series of occurrences. Its standard format is:
repeat(<entity>,<distance>)+<occurrence number>. For instance,
repeat("acgt",[0,3])+[7,38] states that substring *acgt* is repeated from 7 to
38 times, using a spacing of at most 3 characters between 2 repeats.

• **Views and Scope of Constraints:** Constraints (on the content, the size...)
can be set on various parts of a model. They can be imposed to elementary
entities like strings or variables as it has been shown previously, or to a set of
entities that have themselves individually their own constraints.

If the set represents *contiguous elements*, it is called *a view*. In Logol syntax, a
view is delimited by parentheses. In the following example, the model considers
strings built from the concatenation of the instances of 3 variables X1, X2 and
X3, each one having length up to 10 characters. A supplementary constraint on
the view made of the whole string (X1,X2,X3) requires that its total length is
bounded between 8 and 20 characters.

(X1:{#[1,10]}, X2:{#[1,10]}, X3:{#[1,10]}) : {#[8,20]}

It is also possible to set some constraints on a collection of *non-contiguous elements* (for instance on the two segments that form the stem of a stem-loop in a
RNA structure). Such constraints are set in this case in a specific global module,
the *control panel*. The following example details a stem-loop structure made of
two stem elements that have to contain globally at least 30% of *C*.

```
controls:{
% "c"[mod1.ISTEM1,mod1.ISTEM2]>=30
}
mod1()==> STEM1:{#[2,18],_ISTEM1},.*:{#[1,10]}, -"wc" ?ISTEM1:{_ISTEM2}
mod1()==*>SEQ1
```

• **Multiple Analyses:** The coexistence of alternative structures in a same region
is certainly amongst the important features of biological sequences. Gene overlapping for instance has been found in all kingdoms of life, including viruses and

higher eukaryotes. Logol allows to model such situations by stating alternative
models in the grammar top rule (==*> SEQ1). Then, a sequence is accepted only
if it contains an instance of each possible alternative. For instance the grammar:

```
mod1().mod2()==*> SEQ1
mod1() ==> "yvcpfdgcnk"
mod2() ==> "nklkshil"
```

accepts the sequences containing both the strings *yvcpfdgcnk* and *nklkshil*,
independently of their positions, being overlapping or not. Parameter-passing
is possible between alternative models, e.g. to settle the respective positions of
alternative elements.

```
mod1(SAVE1).mod2(SAVE1)==*> SEQ1
mod1(SAVE1) ==> "aata":{_SAVE1},X1:{#[30,30]}:{% "gc":60}
mod2(SAVE1) ==> "gggcaa":{@[@SAVE1 - 20,@SAVE1 + 20]}
```

The above model is looking for an instance of string *aata* that is both followed by
a GC-reach area and contains a neighboring occurrence of string *gggcaa*. Indeed,
@[@SAVE1 - 20,@SAVE1 + 20] constrains the *gggcaa* string to be located 20 nt
before or after the *aata* string.

3   Logol Implementation

3.1   Input /Output Specifications

The Logol software is in charge of matching a Logol pattern against one or
more (DNA, RNA or protein) sequences, in order to point out all the pattern
occurrences within the sequences.

To this end, it needs two main inputs: a Fasta file with the sequences to be
analyzed, and a textual file with the grammatical rules of the pattern. The tool
accepts also a configuration file, setting some parsing parameters. This allows,
among other, to limit the scope of the search by limiting the maximum number
of matches, to choose the indexing tool (Vmatch or Cassiopee, see below), or to
detect and filter irrelevant match variants.

The application outputs a zip archive containing one XML file per input sequence. Output files record all the details of the matches, including the matching
rule, the location of the match, its length, and the number of substitutions and
indel. It also keeps the match information in a tree hierarchy. A repeat, for
example, will be decomposed in an array of matches. Thus, it is possible to analyze the result of the global match, but also the details of any element of the
grammar. In addition to the XML output, with match details, the analyzer can
also output the results in Fasta or GFF3 format. Those formats ease exploiting
results within workflows using other tools.

The core application to launch a Logol analysis is a command-line application,
available for different operating systems. However, there exists also user interfaces, for more comfort. Among them, the *model designer* let the user draw the
model graphically, and the web application converts it into a Logol grammar.

3.2   Sequence Analysis

Pattern matching is performed in two stages. At first the Logol pattern is deciphered by the grammar analyzer, then it is applied on the input sequences by the sequence matcher.

- **Parser:** The grammar analyzer is a Java program. Its role is to decode the grammar to generate a script used by the sequence matcher, and to launch the calls to the sequence matcher. The generated script is a Prolog file which uses a dedicated library containing Prolog predicates for each kind of grammar element (spacers, repeats, ...). The grammar follows a DSL (Domain Specific Language) analyzed by the Antlr library (http://www.antlr.org/). The Prolog programming language has been chosen for its flexibility and conciseness in expressing parsers, due to its built-in ability for backtracking on partial solutions and natural handling of non determinism. However, the implementation could have been achieved with any other language.

To generate the script, several parsing runs are achieved. The first parsing stage gathers information on each element (expected position range, minimum and maximum size, number of allowed errors, ...).

A second parsing stage tries to solve cases where a variable is used but will be instanciated later in the model, e.g in *"acgt",?X1,"cgta":{ _X1 }*. Indeed, though the grammar itself does not require a left to right reading, we use in practice a left to right parsing of the sequence. To manage such cases, the tool finds the variables in this specific situation and applies a dedicated search technique. It also uses information gathered in the first step to add as many constraints as possible on the variable (length, content, ...) in order to reduce the search space.

The last step, using information from previous stages, generates the Prolog script that will be used by the sequence matcher.

Once the script is generated, the analyzer tries to split the input sequence in smaller parts. Indeed, if grammar analysis or input parameters show match length to be smaller than a fixed integer N, then sequences can be cut in several parts (according to configuration, but at least 2N long). This is used to parallelize the search (multi-threading or using a DRMAA compliant cluster). The analyzer triggers sequence matcher runs on each sequence part and merges the results. In case of multiple input sequences, each sequence analysis will also be parallelized.

If multi-thread is used, the program will limit the number of parallel analysis according to the configuration. If DRMAA is used, the program will also try to use multi-thread on the remote node if sequences can be cut in smaller parts.

- **Sequence Matcher:** The sequence matcher is a Prolog compiled script that loads the script generated by the grammar analyzer. It has been tested with Sicstus Prolog (sicstus.sics.se) and SWI-Prolog (swi-prolog.org). It scans the input sequence with the input script rules, trying to match each rule one by one. When a complete rule is matched, it records the match.

For each rule element, the matcher takes a chunk of the sequence and tries to apply the rule on the chunk. If it matches, it goes to the end location of the match and tries to apply the next rule. The matcher records all the details of the match in an XML format. The matcher will optionally apply a filter to delete redundant matches at different levels since it is possible to get two matches at the same location that only differ by their parsing structures.

In case of spacers in a model, the matcher calls an external program using indexing sequence techniques to directly look for positions of the following words. Two possibilities are offered by Logol to perform indexing: VMatch or Cassiopee.

VMatch[1] is a suffix array search tool that supports substitution and indel search. An index is created at startup and the matcher calls the VMatch program to search for a pattern in sequences. VMatch is not open source, but is free for academics. The tool is not delivered with the Logol software suite and needs a manual installation. It is efficient for large sequences.

Cassiopee (`https://github.com/osallou/cassiopee`) is a Ruby tool, developed for *Logol*, though it can be used independently. It scans the sequence to match a pattern with error support. This tool has been developed to provide a complete open source solution, but it is not as efficient as VMatch for large sequences. Then, VMatch usage is the recommended choice when performance is crucial. The tool selection is made in the configuration file so that it can be adapted for each analysis.

4   Illustration: Modeling -1 Ribosomal Frameshifting

RNA recoding is a fundamental biological mechanism that cells use to expand the number of proteins assembled from a single DNA code. There are several types of RNA editing modifying the standard translation of a messenger RNA by the ribosome, which seem largely directed by the 3D conformation adopted by the RNA molecule. The *-1 programmed ribosomal frameshifting* (PRF) is a recoding event which occurs when the ribosome is moving rearward exactly 1 nt on a 'slippery site', X XXY YYZ, where X,Y and Z are nucleotides. The ribosome reads the first X nucleotide two times in this case. Indeed, while standard translation processes codons ABX XXY YYZ CDE ..., the codons processed by PRF are ABX XXX YYY ZCD .... The typical structure promoting a -1 frameshifting event, which we call in the following "PRF pattern", is sketched in Figure 1. It is made sequentially of a start codon, a number of codons, a heptameric slippery site XXXYYYZ placed in -1 phase, a few nucleotide spacer, and a characteristic stable secondary structure. The secondary structure is the obstacle stopping the ribosome during the heptamer translation and triggering a movement one nucleotide backwards that causes reading frameshifting. The secondary structure is usually made of a "H-type pseudo-knot" including two nested stem-loops.

A number of tools exists for the detection of putative sites where a -1 frameshifting event might occur [7], but this detection process remains an active research topic since the PRF pattern is not universal (the characteristic features of the heptamer, the spacer, and the secondary structure depend on the organism) and the detection of pseudo-knots is a difficult issue. Many methods proceed by successive filtering steps like KnotInFrame[21], one of the most advanced tool in this category. KnotInFrame initially detects all heptamers XXXYYYZ, then

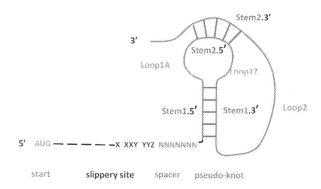

Fig. 1. "PRF pattern"= typical structure promoting a -1 frameshifting event

looks for potential pseudo-knots downstream of this motif, using a dedicated RNA folding procedure.

4.1   PRF Logol Model

The complexity of the PRF pattern makes it a good candidate to investigate the expressivity of Logol. In order to elaborate the corresponding model, one needs to use a number of Logol features like multi-analysis, negative content constraints, repeated motifs or search for biological palindromic structures. We present here the most prominent aspects of the model.

• **Multi-analysis of Two Overlapping ORFs:** Among the mandatory structural features for the occurrence of a -1 frameshifting event, some are concerning the reading frame setting. The standard translation occurs in a sequence with an open reading frame: a start codon (AUG) followed by a number of non stop codons (triplets), and a stop codon (UGA,UAG ou UAA) terminating the translation. All these codons are *in 0 phase*. The alternative translation, in case of -1 frameshifting, starts on the same start codon but moves backward one nucleotide on the slippery site, leading to proceed further on triplets *in -1 phase* until a stop codon is reached, also in -1 phase.

In order to possibly generate a -1 frameshifting event, a RNA sequence should thus contain both an open reading frame in 0 phase (a start followed by a sufficient numbers of codons ending by a stop codon) and at a constrained distance from the start, a series of codons ending by a stop in -1 phase.

This check is triggered in Logol by a *multiple analysis* recognizing alternative patterns, "ORF" and "ORFminus", on the same string with a *parameter-passing between the two models* in order to share the common start. The ORF model thus contains a *repeat* that accepts up to 300 non stop codons (a value consistent with the literature), that is: repeat(notstop(),[0,0])+[0,300], where notstop stands for a model built from a *view*, that accepts a string of length 3 that is not a stop. This is achieved by a *negative content constraint* on the view.

• **Slippery Site and Spacer:** The PRF pattern describes 3 segments: the slippery site, the spacer and the secondary structure. The Logol model for the *slippery site* respects the consensus: it is an heptameric motif in the form XXXYYYZ, which must be positioned in -1 phase, where X is any nucleotide repeated 3 times, Y is the base A or U repeated 3 times, and Z differs from base G, according to the Logol pattern mod3 below.

```
mod3()==> mod4(), (("aaa")|("uuu")), ! "g":{#[1,1]}
mod4()==> (("aaa")|("ccc")|("uuu")|("ggg"))
```

The *spacer* is straightforwardly described by a gap element (of size less than 10 in this case).

• **Pseudo-knots:** The most efficient secondary structure for -1 frameshifts is the pseudo-knot of type H (two interwined stem-loops, cf 1), even if it is not the sole existing structure. It is thus the structure that has been modeled here.

We first provide a simplified Logol grammar for pseudo-knot structures. In this grammar, STEM15 refers to the first strand (in the 5' direction) of the first stem and -"wc" ?IS15 refers to its 2nd strand (in the 3' direction), which is its reverse complement up to 4 mismatches. STEM25 and -"wc" ?IS25 refer to the two elements of the 2nd stem. The gaps refer to the loop elements between stems.

```
STEM15:{#[4,16],_IS15},.*:{#[1,5]}, STEM25:{#[3,8],_IS25},.*:{#[0,4]},
  -"wc" ?IS15 :{$[0,4]},.*:{#[4,40]},-"wc" ?IS25 :{$[0,2]}
```

Our validation process on real data (next paragraph) resulted in a significant refinement of this model (cf Figure 2). The final model makes use of a great variety of Logol language elements. Among new elements, the model integrates the count of the GC ratio in stems, the separate treatment of nucleotides at the end of stems in order to forbid mismatches at these positions, or the possibility of non-canonical Wobble pairing (G-U), called wcw here, at particular stem positions [17]. An excerpt from the Logol grammar dedicated to the first stem follows, the whole model being presented in figure 2:

```
// 50% of GC pairing in Stem1 => 25% of C in [Stem1.5' + Stem1.3']
controls:{ % "c"[mod5.IA5,mod5.IS15,mod5.IZ5,mod5.IZ3,mod5.S13,mod5.IA3]>=25}
mod5()==> (A5:{#[1,1],_IA5},S15:{#[2,14],_IS15},Z5:{#[1,1],_IZ5}):{%"gc":50},
    LOOP1:{#[1,5]},        ... stuff deleted ....
 -"wcw" ?IZ5:{_IZ3}, -"wc" ?IS15:{_S13}:{p$[0,34]}, -"wcw" ?IA5:{_IA3}, ...
```

• **A First Model Validation:** In order to test and refine our -1 frameshift model, we have elaborated a sequence test set [17] around sequences known to produce -1 frameshift events. Thirty proven sequences ("validated -1 frameshift") from the reference base Recode2 (recode.genetics.utah.edu) have been completed by random sequences obtained by shuffling 100 times each reference sequence, using Shuffleseq (emboss.bioinformatics.nl/cgi-bin/emboss/). This procedure keeps the sequence lengths and nucleotidic ratios of the reference. The final set thus contains 30 "positive" sequences (in which one expects to find the

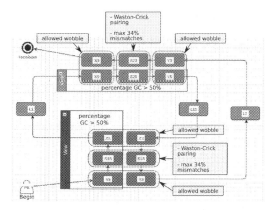

Fig. 2. Global overview of the final Logol model for the PRF pseudo-knot

PRF pattern at the right position) and 300 "negative" sequences (in which one expects a minimum number of PRF pattern occurrences).

This validation work [17] led us to perform comparisons between the Logol prediction of pseudo-knots and those made by "DotKnot", a pseudo-knot prediction software using RNAfold to compute the probability of each folding structure (http://dotknot.csse.uwa.edu.au). It brought about some significant changes in our initial model: parameter tuning has concerned wooble pairing, GC ratio and the mismatch rate in the stems. Ultimately, the Logol model finds about 100 matches per sequence on the Recode2 reference set, among them being in most cases the desired match. It is possible to compute a posteriori a quality score for each stem and sort accordingly the matches[17]. This procedure leads to the right frameshift area prediction in 20 cases over the 30. The score is based on a pairing cost function proposed by J.P. Forest [8] for the stem: {GCpairing=+3, AUpairing=+2, GUpairing=+1, mismatch=-2}.

- **Runtime.** To give an idea of the performances, parsing the largest reference sequence (30Kb) with the final Logol model takes 1'30s (on a PC Intel X5550, 144Go RAM), while the KnotinFrame answer is immediate on such a sequence. The complete analysis of the Bacillus subtilis genomic sequence (*str168NC_000964.3*, 4.2 Mbp) produces 7000 matches in 2 hours. KnotinFrame web site does not accept such a large sequence.

5   Conclusion

The Logol pattern matching tool has been conceived to allow the modeling and search of realistic structures in biological sequences. It has been designed to be expressive but also evolutive, in order to ease the introduction of new features. The fact that the language has proved fairly well suited to model the complex pattern of ribosomal frameshift, whereas it was not designed for this task, seems an encouraging sign on the genericity of the language elements. The tool is operational and available on the GenOuest bioinformatics platform, under CeCILL license. Although efforts have been made to offer a wide access to Logol functionalities through either command-line or graphical interface inputs, we welcome any user feed-back to increase its ergonomic features and actual range of applicability.

References

1. Abouelhoda, M.I., Kurtz, S., Ohlebusch, E.: Replacing suffix trees with enhanced suffix arrays. J. Discrete Algorithms 2(1), 53–86 (2004)
2. Billoud, B., Kontic, M., Viari, A.: Palingol: a declarative programming language to describe nucleic acids' secondary structures and to scan sequence database. Nucleic Acids Res. 24(8) (1996)
3. de Castro, E., Sigrist, C.J.A., et al.: Scanprosite: detection of prosite signature matches and prorule-associated functional and structural residues in proteins. Nucleic Acids Research 34(suppl. 2), 362–365 (2006)
4. Dong, S., Searls, D.B.: Gene structure prediction by linguistic methods. Genomics 23(3), 540–551 (1994)
5. Dsouza, M., Larsen, N., Overbeek, R.: Searching for patterns in genomic data. Trends in Genetics 13(12), 497–498 (1997)
6. Eddy, S.: Rnabob: a program to search for rna secondary structure motifs in sequence databases (1996)
7. Firth, A.E., Bekaert, M., Baranov, P.V.: Computational resources for studying recoding. In: Atkins, J.F., Gesteland, R.F. (eds.) Recoding: Expansion of Decoding Rules Enriches Gene Expression, Nucleic Acids and Molecular Biology, vol. 24, pp. 435–461. Springer, New York (2010)
8. Forest, J.P.: Modélisation et détection automatique de sites de décalage de cadre en -1 dans les génomes eucaryotes. Ph.D. thesis, Université de Paris VI (2005)
9. Gattiker, A., Gasteiger, E., Bairoch, A.: Scanprosite: a reference implementation of a prosite scanning tool. Applied Bioinformatics 1(2), 107–108 (2002)
10. Graf, S., Strothmann, D., Kurtz, S., Steger, G.: HyPaLib: a Database of RNAs and RNA Structural Elements defined by Hybrid Patterns. Nucleic Acids Res. 29(1), 196–198 (2001)
11. Jensen, K., Stephanopoulos, G., Rigoutsos, I.: Biogrep: A multi-threaded pattern matcher for large pattern sets (2002)
12. Joshi, A.K., Vijay-Shanker, K., Weir, D.: The convergence of midly context-sensitive grammars. In: Shieber, S.M., Wasow, T. (eds.) The Processing of Natural Language Structure, pp. 31–81. MIT Press, Bosto (1991)
13. Macke, T.J., Ecker, D.J., Gutell, R.R., Gautheret, D., Case, D.A., Sampath, R.: Rnamotif, an rna secondary structure definition and search algorithm. Nucleic Acids Research 29(22), 4724–4735 (2001)
14. Meyer, F., Kurtz, S., et al.: Structator: fast index-based search for rna sequence-structure patterns. BMC Bioinformatics 12(1), 214 (2011)
15. Nicolas, J., Durand, P., et al.: Suffix-tree analyser (stan): looking for nucleotidic and peptidic patterns in chromosomes. Bioinformatics 21(24), 4408–4410 (2005)
16. Pesole, G., Liuni, S., DSouza, M.: Patsearch: a pattern matcher software that finds functional elements in nucleotide and protein sequences and assesses their statistical significance. Bioinformatics 16(5), 439–450 (2000)

17. Rocheteau, A., Belleannée, C.: Recherche d'éléments structurés dans les génomes par modèles logiques. Rapport de recherche PI-1994, Dyliss - Inria - Irisa (April 2012), http://hal.inria.fr/hal-00684388
18. Searls, D.B.: String variable grammar: A logic grammar formalism for the biological language of DNA. Journal of Logic Programming 24(1&2), 73–102 (1995)
19. Searls, D.B., Dong, S.: A syntactic pattern recognition system for DNA sequences. In: Cantor, C.R., Lim, H.A., Fickett, J., Robbins, R.J. (eds.) Proceedings 2nd International Conference on Bioinformatics, Supercomputing, and Complex Genome Analysis, pp. 89–101. World Scientific, Singapore (1993)
20. Strothmann, D., Gräf, S.A., Kurtz, S., Steger, G.: The syntax and semantics of a language for describing complex patterns in biological sequences. Tech. rep., Universität Bielefeld, Arbeitsgruppe Praktische Informatik (August 2000)
21. Theis, C., Reeder, J., Giegerich, R.: Knotinframe: prediction of -1 ribosomal frameshift events. Nucleic Acids Research 36(18), 6013–6020 (2008)

Evolutionary Algorithm Based on New Crossover for the Biclustering of Gene Expression Data

Ons Maâtouk[1,2], Wassim Ayadi[2,3], Hend Bouziri[1], and Beatrice Duval[2]

[1] LARODEC Laboratory, ISG Tunis, Université de Tunis, Tunisia
[2] LERIA, Université d'Angers, 2 Boulevard Lavoisier, 49045 Angers, France
[3] LaTICE Laboratory, ESSTT, Université de Tunis, Tunisia
mtk-ons@hotmail.fr, {wassim.ayadi,hend.bouziri}@gmail.com,
bd@info.univ-angers.fr

Abstract. Microarray represents a recent multidisciplinary technology. It measures the expression levels of several genes under different biological conditions, which allows to generate multiple data. These data can be analyzed through biclustering method to determinate groups of genes presenting a similar behavior under specific groups of conditions.

This paper proposes a new evolutionary algorithm based on a new crossover method, dedicated to the biclustering of gene expression data. This proposed crossover method ensures the creation of new biclusters with better quality. To evaluate its performance, an experimental study was done on real microarray datasets. These experimentations show that our algorithm extracts high quality biclusters with highly correlated genes that are particularly involved in specific ontology structure.

Keywords: Biclustering, Evolutionary algorithm, Crossover method, Microarray data, Data mining.

1   Introduction

During recent years, microarray technology has reached a main role in biological and biomedical research [24]. This technology measures the expression levels of thousands of genes in different biological conditions. It allows to generate large amount of data [14]. The analysis of these data allows the extraction of biological knowledge in order to understand diseases [4]. Given the huge masses of data to be analyzed, the use of data mining techniques has become essential to extract the knowledge embedded in these masses of information. Among the clustering techniques, we can find the biclustering which has been used extensively to analyse gene expression data.

The biclustering is a data mining technique to discover high quality biclusters. These biclusters are illustrated by groups of genes presenting a similar behavior under specific groups of conditions. Formally, the biclustering problem [26] is to build a group of biclusters associated with a data matrix taking account a fitness

M. Comin et al. (Eds.): PRIB 2014, LNBI 8626, pp. 48–59, 2014.
© Springer International Publishing Switzerland 2014

function that measures the quality of a group of biclusters. Thus, it is highly combinatorial problem [26] and known to be NP-Hard [5].

Given the robustness to dynamic changes of the evolutionary approach and their ability to self-optimization, we adopt this approach to solve the biclustering problem. Most of the biclustering algorithms based on the evolutionary approach, like [1,8,9], use random crossover method. However, these methods do not guarantee to obtain a better quality child biclusters, that prompt us to seek a crossover method specific for the biclustering of gene expression data and allowing to have better quality biclusters.

In this work, we propose an *Evolutionary Biclustering Algorithm based on a new Crossover method* (EBACross). This new method is dedicated to the biclustering of gene expression data. EBACross uses a fast local search algorithm to generate an initial population with reasonable quality. A selection and a mutation operator are used. After an experimental study, we notice that our proposed algorithm can extract high quality biclusters with highly correlated genes which are particularly involved in specific ontology structure.

2   Description of the Proposed Biclustering Algorithm

In order to extract high-quality biclusters, we propose a new biclustering algorithm adopting the evolutionary approach. It can be summarized by 5 steps:

1. Generate the initial population P_{init}. This step is based on the Cheng and Church algorithm [5]. It is recognized for its reasonable results in a quick time and its almost total coverage of genes and conditions. It allows to start with reasonable quality biclusters covering almost all the data matrix.
2. Build the parent set P by selecting the best biclusters of the initial population P_{init}. The selection is based on four complementary fitness functions: size (f_1), MSR (f_2), average correlation (f_3) and coefficient of variation (f_3).
3. Create children biclusters by our proposed crossover operator that is dedicated for the biclustering of gene expression data. Based on a discretization method and the standard deviation function, this crossover combines the biclusters parents in pair giving priority to the biclusters that satisfy a maximum number of fitness functions.
4. In order to avoid overlapping biclusters and to increase the diversification of biclusters a mutation operator is used. This operator is applied to the biclusters resulting to the crossover. It is based on the average correlation function which allows to improve the coherence of gene biclusters.
5. Replace bicluster parents by those resulting to the mutation and repeat from step 2 until the reaching of the number of iterations.

2.1   Biclusters Encoding

To represent the biclusters, the majority of existing biclustering algorithms uses a fixed size binary string [17,19]. This string is built by two bit strings. The first one represents the genes and the second represents the conditions. The string position of a gene (respectively a condition) takes 1 if the gene (respectively the condition) belongs to the bicluster, 0 otherwise. This method explores all genes and conditions. It leads to high consumption of time and memory space.

To remedy, we represent biclusters as string composed by an ordered gene and condition indices like in [7,22,1].

2.2   Selection

The selection method is applied on the initial population P_{init} to build the parent set P. This set includes the best biclusters of P_{init} according the fitness functions. To extract maximal high-quality biclusters of highly correlated genes, we can consider four main complementary fitness functions:

Size: Most of biclustering algorithm defined the size of a bicluster by its number of elements $|G| * |C|$ as in [11]. This function gives more chance to the number of genes to be maximized since the total number of genes is higher than the number of conditions. To be able to choose if we want to give more chance to the number of genes or to the number of conditions to be maximized, we define the size of biclusters by the following function where α and β are two constants.

$$f_1(Bic) = \alpha \frac{|G'|}{|G|} + \beta \frac{|C'|}{|C|} \tag{1}$$

Mean Squared Residue: Cheng and Church [5] proposed *Mean Squared Residue* (MSR) which measures the correlation of a bicluster. A high value of MSR indicates that the bicluster is weakly coherent while a low value of MSR indicates that it is highly coherent. It is defined as follows:

$$f_2(Bic) = \frac{1}{|G'||C'|} \sum_{i \in G', j \in C'} (m_{ij} - m_{iC'} - m_{G'j} + m_{G'C'})^2 \tag{2}$$

where $m_{iC'}$ (respectively $m_{G'j}$) represents the expression level average of the ith row (respectively the jth column), $m_{G'C'}$ corresponds to the expression level average of the bicluster $Bic(G', C')$ and m_{ij} represents the expression level corresponding to the ith row and the jth column.

Average Correlation: Nepomuceno *et al.* [18] proposed the average correlation function to evaluate the correlation between genes in each biclusters. They indicate that the proposed function can find biclusters that cannot be found by the algorithms based on MSR. Due to these, algorithms might not find scaling patterns when the variance of gene value is high. The average correlation of the bicluster $Bic(G', C')$ is defined as follows:

$$f_3(Bic) = \frac{2}{|G'|.(|G'| - 1)} \sum_{i=1}^{G'} \sum_{j=i+1}^{G'} \left| \frac{cov(g_i, g_j)}{\sigma_{g_i} \sigma_{g_j}} \right| \tag{3}$$

where $cov(g_i, g_j)$ represents the covariance of the rows corresponding to the gene g_i and the gene g_j and σ_{g_i} (respectively σ_{g_j}) corresponds to the standard deviations of the rows corresponding to the gene g_i (respectively the gene g_j).

This measure varies between 0 and 1. If the genes are highly correlated, $f_3(Bic) = 1$, 0 otherwise.

Coefficient of Variation: Statistically, the Coefficient of Variation (CV) is used to characterize the variability of the data in a sample by evaluating the percentage of variation relative to its average. The higher the value of the coefficient of variation is, the larger is the dispersion around the average. It allows to compare the variability of several samples that have different average or even which are not expressed in the same units.

By adopting it to the biclustering of microarray data, the coefficient of variation can be considered as a measure to evaluate the variability of genes of a bicluster under all its conditions. This measure is calculated separately for each bicluster and is defined as follows:

$$f_4(Bic) = \frac{\sigma_{Bic}}{m_{G'C'}} \qquad (4)$$

where σ_{Bic} represents the standard deviation of the bicluster Bic and $m_{G'C'}$ corresponds to the average of all the expression levels of the bicluster Bic.

A bicluster with a high coefficient of variation is a bicluster whose the dispersion of expression levels is high. When a bicluster have a coefficient of variation equal to 0 then it has constant values.

So, the parent set P can be divided into four subsets:

- P_1: biclusters from P_{init}, with a value of f_1 higher than the threshold Th_1.
- P_2: biclusters from $P_{init} \setminus P_1$, with a value of f_2 lower than the threshold Th_2.
- P_3: biclusters from $P_{init} \setminus (P_1 \cup P_2)$, with a value of f_3 higher than the threshold Th_3.
- P_4: biclusters from $P_{init} \setminus (P_1 \cup P_2 \cup P_3)$, with a value of f_4 higher than the threshold Th_4.

2.3 Crossover

In order to obtain children biclusters with a better quality than their parent biclusters, we propose a new crossover method specific for the biclustering of gene expression data. Unlike the random crossover method used for the biclustering of gene expression data [1,22], our crossover considers the two parts of bicluster (genes and conditions part) simultaneously. It is essentially based on five steps :

Selection of the Biclusters to Combine: It consists to select the biclusters which satisfies more fitness functions to combine them together.

Let's consider the four bicluster parents : Bic_0, Bic_1, Bic_2 and Bic_3. Table 1 represents the satisfaction of the four parent biclusters to the different fitness functions. Bic_1 satisfies all the fitness functions, Bic_2 satisfies three fitness functions while Bic_0 and Bic_3 satisfy only two. So, we start by combining the biclusters Bic_1 and Bic_2. Then, we combine the biclusters Bic_0 and Bic_3.

Table 1. Satisfaction of the biclusters to the different fitness functions

Biclusters	f_1	f_2	f_3	f_4
Bic_0	$< Th_1$	$< Th_2$	$> Th_3$	$< Th_4$
Bic_1	$= Th_1$	$< Th_2$	$> Th_3$	$> Th_4$
Bic_2	$< Th_1$	$< Th_2$	$> Th_3$	$> Th_4$
Bic_3	$< Th_1$	$> Th_2$	$< Th_3$	$= Th_4$

Creation of the Total Bicluster: This step consists to merge the sets of genes G_1 and G_2 (respectively the sets of conditions C_1 and C_2) of the two parent biclusters Bic_1 and Bic_2 into a single set G (respectively C). This allows to create a new bicluster Bic_{Tot}.

Let's consider the following bicluster parents :

$$Bic_1 = (\ |G_1| \times |C_1|\ )$$
$$Bic_2 = (\ |G_2| \times |C_2|\ )$$

where :
$G_1 = \{\ g_1, g_2, \ldots, g_n\ \}$ and $G_2 = \{\ g'_1, g'_2, \ldots, g'_m\ \}$ correspond respectively to the sets of genes of the two parent biclusters Bic_1 and Bic_2.
$C_1 = \{\ c_1, c_2, \ldots, c_p\ \}$ and $C_2 = \{\ c'_1, c'_2, \ldots, c'_q\ \}$ correspond respectively to the sets of conditions of the two parent biclusters Bic_1 and Bic_2.

The merge of these two biclusters gives a bicluster $Bic_{Tot} = (\ |G| \times |C|\ )$ where $G = G_1 \cup G_2$ corresponds to the set of genes and $C = C_1 \cup C_2$ corresponds to the set of conditions.

Discretization of the Total Bicluster: To cluster the conditions with similar expression levels for each gene, a discretization method is applied independently for each one. This requires the decomposition of the total bicluster into several vectors. Each vector represents the expression levels of a specific gene under all conditions of total bicluster. The discretization method is based on the *Standard Deviation* (SD) to determine whether conditions can belong to the same cluster. It is a statistical measure to evaluate the dispersion of a value around the average. This measure is defined as follows:

$$SD_C = \sqrt{\frac{1}{t-1} \sum_{i=1}^{t} (a_i - \bar{C})} \tag{5}$$

where $C = \{c_1, c_2, \ldots, c_t\}$ represents a set of t conditions, a_i represents the expression level of the ith condition and $\bar{C}$ represents the average of the expression levels of the set C, for a specific gene.

The discretization method can be summarized by the following steps:

1. Sort the set C, according to their expression levels in ascending order, to construct a new conditions set $C' = \{c'_1, c'_2, ..., c'_t\}$. This part is important. It ensures a better clustering and optimal clusters for the next part.
2. Cluster the conditions of the set C' based on the standard deviation. First, calculate the standard deviation SD_C of the vector. Then, check whether the cluster Cl is empty and browse the conditions one by one.
3. If $Cl = \emptyset$, add the condition c'_j to Cl and return to the previous step. Else, add also the condition c'_j to Cl and calculate its standard deviation SD_{Cl}.
4. If $SD_{Cl} > SD_C$ or $SD_{Cl} > SD_{Old}$ (SD_{Old} standard deviation of Cl before adding the last condition c'_j), assign c'_j to the next cluster and return to the step 2. Otherwise, to be sure that the condition c'_j is closer to the conditions of Cl than the condition following c'_{j+1}, calculate the standard deviation SD_{Next} of c'_j and c'_{j+1}.
5. If ($SD_{Cl} <= SD_{Next}$), assign the condition c'_j to the next cluster and repeat all the steps. Otherwise, let the condition c'_j in this cluster Cl, return to the step 2 and repeat with the condition c'_{j+1}.

These steps are repeated for each vector. Once complete, return vectors in their original order. Then, bring them together to construct the discretized matrix $Disc$. The cell of this matrix D_{ij} indicates the index of the cluster to which the jth condition belongs for the ith gene.

Construction of the Variation Matrix: Based on the discretized matrix, we build a new matrix. This matrix shows the variation of genes between each pair of conditions. Columns represent genes and rows represent the pair of conditions $\{(c_1\text{-}c_2), (c_1\text{-}c_3), (c_1\text{-}c_3) ..., (c_{t-1}\text{-}c_t)\}$. The cells of the matrix V_{ij} can take only three values:

- If $D_{ia} > D_{ib} : V_{ij} = \text{-}1$ with $a \leq t$, $b \leq t$ and $a < b$
  For the ith gene, the index of the cluster to which belongs the condition a is higher than the cluster to which belongs condition b.
- If $D_{ia} = D_{ib} : V_{ij} = 0$ with $a \leq t$, $b \leq t$ and $a < b$
  For the i_{th} gene, both conditions a and b belong to the same cluster.
- If $D_{ia} < D_{ib} : V_{ij} = 1$ with $a \leq t$, $b \leq t$ and $a < b$
  For the ith gene, the index of the cluster to which belongs the condition a is lower than the cluster to which belongs condition b.

Search of Children Bicluster: The last step allows to extract the biclusters by browsing variation matrix and selecting genes having the same index.

Let's consider the example in Table 2. First, check if there are other genes with the same index as the gene g_0 for the pair of conditions $(c_0\text{-}c_1)$. Only the gene g_3 is found. Now, check if these two genes g_0 and g_3 have the same index for other pairs of conditions. In this example, the genes g_0 and g_3 have the same index for all pairs of conditions. So, the first child bicluster contains the genes g_0, g_3 and the conditions c_0, c_1, c_2, c_3 ($Child_1 : 0\ 3\ //\ 0\ 1\ 2\ 3$).

Then, do the same steps with the gene g_1. The index of the gene is different from all other genes for the conditions, as well as for the gene g_2. Therefore, back to the gene g_0 for the pair of conditions (c_0-c_2) and check if there are any other gene with the same index. Only the gene g_3 is found. It is the case of the first child. Thus, ignore it and go to the next gene. So on, until finding all children biclusters. To avoid overlapping biclusters, we used the *Jaccard index* [13]. This index measures the overlap between two biclusters in terms of genes and conditions.

Table 2. Example of variation matrix

	c_0-c_1	c_0-c_2	c_0-c_3	c_0-c_4	c_1-c_2	c_1-c_3	c_1-c_4	c_2-c_3	c_2-c_4	c_3-c_4
g_0	-1	-1	0	1	0	1	1	1	1	1
g_1	1	0	0	1	-1	-1	0	0	1	1
g_2	0	1	1	1	1	1	1	0	1	1
g_3	-1	-1	0	1	0	1	1	1	1	1

This crossover method allows to create children biclusters with a better quality. The use of standard deviation to discretize parent biclusters allows to group closest expression levels for each gene and to construct the variation matrix. This matrix indicates the variation of the expression levels between each pair of conditions, which allows to determinate the genes presenting a similar behavior and to extract biclusters with highly correlated genes.

2.4   Mutation

In order to ensure the diversification of biclusters and to improve their quality, a mutation method is applied. It tries to improve the coherence between the genes of the biclusters obtained from the crossover, using a correlation matrix. This genetic operator seeks the less coherent gene in the bicluster. Then, it replaces this less coherent gene by the most coherent gene which does not belong to the bicluster.

To construct the correlation matrix, we must calculate the correlation coefficient between each pair of genes. Then, depending on the value obtained, a value is assigned to the cell C_{ij} corresponding to the correlation between the gene g_i and the gene g_j. The cell C_{ij} can take only three values: $C_{ij} = -1$, if $i = j$. Otherwise, $C_{ij} = 0$ when $\left|\rho_{(g_i,g_j)}\right| < Th_{Corr}$ and $C_{ij} = 1$, when $\left|\rho_{(g_i,g_j)}\right| \geq Th_{Corr}$.

3   Experimental Results

In order to test the performance of our proposed algorithm and analyze its results, a series of experiments is performed on real gene expression datasets: *Yeast cell cycle* [25] and *Saccharomyces cerevisiae* [10]. The evaluation of biclustering algorithms and its comparison are based on two complementary criteria: statistical criteria and biological criteria. We compare the results of EBACross with other sate-of-the-art biclustering algorithms ISA [3], BiMax [20], CC [5], OPSM [2], X-Motif [16] and the evolutionary algorithm EvoBic [1], H-MOBI [22], SEBI [8].

3.1 Statistical results

To measure the quality of resulting biclusters, we use the functions "size", "average correlation", "MSR" like in [1,18,22]. We calculate the "coverage" and we proceed as in [9,11,15]. This criterion is defined as being the total number of cells of the matrix M covered by the resulting biclusters.

Table 3. Comparing the fitness function values and the coverage of different biclustering algorithms for the *Yeast Cell Cycle* and *Saccharomyces cerevisiae* datasets

	||EvoBic|	H-MOBI|	SEBI	|BiMax|	CC	ISA	|OPSM|	X-Motif|	EBACross|
			Yeast Cell Cycle						
Gene number	788,4	1610,8	—	24,0	39,62	76,3	437,94	1,2	38,08
Condition number	3,3	7,87	—	3	3,16	8,7	9,5	11,4	3,78
Average correlation	0,90	—	—	0,66	0,84	0,50	0,91	0,71	0,82
MSR	291	297	205,18	209,5	10,94	248,25	288,04	203,14	167,62
Genes coverage	99,58	—	13,61	79,09	61,79	73,44	83,98	52,86	66,85
Conditions coverage	70,59	—	15,25	64,71	100	100	100	100	100
Total coverage	44,21	—	—	46,48	10,75	38,94	18,67	25,36	49,53

	||EvoBic|	H-MOBI|	SEBI	|BiMax|	CC	ISA	|OPSM|	X-Motif|	EBACross|
			Saccharomyces cerevisiae						
Gene number	17,8	—	—	32,8	81,11	76,27	95,58	1,12	41,46
Condition number	3	—	—	3	19,64	8,71	12,5	34,52	4,20
Average correlation	0,90	—	—	0,68	0,33	0,59	0,87	0,97	0,81
MSR	0,08	—	—	0,18	0,36	0,22	0,08	10^{17}	0,25
Genes coverage	4,84	—	—	29,54	96,06	34,08	13,79	10,89	85,77
Conditions coverage	7,51	—	—	79,19	100	58,38	26,01	100	84,39
Total coverage	0,09	—	—	0,99	49,09	2,25	0,96	2,62	10,12

TABLE 3 presents the average value of the gene number, the condition number, the average correlation, the MSR and the coverage of the obtained biclusters for the *Yeast Cell Cycle* and *Saccharomyces cerevisiae* datasets.

We can show that most algorithms have relatively close results. For the *Yeast Cell Cycle*, the best MSR value is obtained by CC (MSR = 10,94) and the best average correlation value is obtained by OPSM (ρ = 0,91) while for the *Saccharomyces cerevisiae* dataset, the best MSR and average correlation value are obtained by X-Motif (MSR = 10^{-17} and ρ = 0,97). Although the results of our proposed algorithm are not the best, they have a satisfactory quality and are consistent. Indeed, we note an average correlation value equal to 0,82 (respectively 0,81) and a MSR value equal to 167,62 (respectively 0,25) for the *Yeast Cell Cycle* (respectively the *Saccharomyces cerevisiae*) dataset.

Concerning the percentage of cells in the initial matrix covered by the different biclustering, we can show that most algorithms have relatively close results.

However, our algorithm has the best percentage for the genes coverage, conditions coverage and total coverage. Indeed, for the *Yeast Cell Cycle* (respectively the *Saccharomyces cerevisiae*) dataset, the biclusters generated by our algorithm cover 66,85% (respectively 85,77%) of the genes, 100% (respectively 84,39%) of the conditions and 49,53% (respectively 10,12%) of the cells of the initial matrix.

3.2 Biological Results

The biological criteria allows to measure the quality of resulting biclusters, by checking whether the genes of a bicluster have common biological characteristics. For that, we calculate the p-value. The biclusters with a p-value p lower than 5% are considered as over-represented. The most obtained biclusters have a p-value close to 0, i.e., the most genes of this bicluster have common biological characteristics.

Given the large number of the obtained biclusters, We proceed as in [8,12,20] and we test only on the one hundred best biclusters.

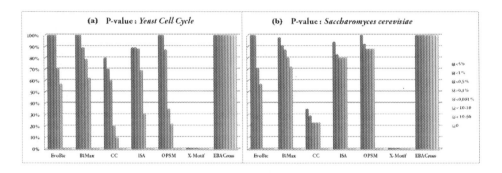

Fig. 1. P-value : *Yeast Cell Cycle* and (a) *Saccharomyces cerevisiae* dataset (b)

FIGURE 1 show the percentage of extracted biclusters for different adjusted p-value (p = 5%; 1%; 0,5%; 0,1%; 0,001%; 10^{-10}; 10^{-50} and 0), for the *Yeast Cell Cycle* and the *Saccharomyces cerevisiae* datasets.

We can show that the majority of the algorithms have rather low percentage. For the *Saccharomyces cerevisiae* dataset, 72%, 80% and 88% of the biclusters respectively extracted by Bimax, ISA and OPSM are statistically significant with a p-value $p < 0,001\%$. Only EBACross reaches a value less than 10^{-10}. Indeed, 100% of the biclusters extracted by our algorithm are statistically significant with a p-value equal to 0. However, for the *Yeast Cell Cycle* dataset, we notice a degradation in the results of the majority of algorithms while our algorithm maintains the quality of its results. Only 62%, 31% and 22% of the biclusters respectively extracted by Bimax, ISA and OPSM are statistically significant with a p-value $p < 0,001\%$ and 100% of the biclusters extracted by EBACross are statistically significant with a p-value equal to 0.

We evaluate also qualitatively the capacity of the algorithms to extract significant biclusters with a biological point of view. It requires the incorporation of biological knowledge. The biological signification of the obtained biclusters can be interpreted based on *Gene Ontology* (GO) [6] for the description of the roles of genes and their products [21]. There are three ontology structures describing the gene products: biological process, molecular function and cellular component.

Given the large number of the obtained biclusters, we proceed as in [19,23] and we present the most significant GO shared of three random biclusters extracted by our algorithm in Table 4 and Table 5, respectively for the *Yeast Cell Cycle* and *Saccharomyces cerevisiae* datasets. These tables include the gene number, condition number and the different shared GO terms for each ontology structures of the biclusters.

Table 4. GO terms of biclusters extracted by EBACross for *Yeast Cell Cycle* dataset

Biclusters	Cellular component	Molecular function	Biological process
3 genes; 9 conditions			*cellular process* (66,7%; 0)
4 genes; 7 conditions		*transferase activity* (75%; 0)	
13 genes; 5 conditions	*nucleolus* (86,7%; $9,4.10^{-32}$)		

Table 5. GO terms of biclusters extracted by EBACross for *Saccharomyces cerevisiae* dataset

Biclusters	Cellular component	Molecular function	Biological process
13 genes; 4 conditions			*regulation of mitotic cell cycle* (38,4%; 0)
4 genes; 5 conditions		*binding* (50%; 0)	
60 genes; 7 conditions	*intracellular* (31,67%; $5,2.10^{-6}$)		

We can show that the extracted biclusters are biologically relevant according to a single ontology structure and for this structure, we find only one GO term. For example, in Table 4, the genes of the first bicluster are particularly involved in the *cellular process* with a p-value $p = 0$, those of the second bicluster are particularly involved in the *transferase activity* function with a p-value $p = 0$, those of the third bicluster are particularly involved in the *nucleolus* component with a p-value $p = 9, 4.10^{-6}$.

We can note that EBACross is efficient to extract signicant biclusters with specific GO term for all ontology structures (biological process, molecular function and cellular component).

4   Conclusion

In this paper, we introduce a new evolutionary algorithm. The selection operator allows to keep the best quality biclusters, based on four complementary functions. Then, a new crossover operator dedicated for the biclustering of gene expression data is used. This proposed crossover ensures the creation of new biclusters with better quality. Finally, based on the average correlation function a mutation operator seeks the least coherent gene in each bicluster to replace it with a more coherent gene.

To evaluate the performance of our algorithm, an experimental study was done on the real microarray datasets. We compare the results to other existing biclustering algorithms. These experimentations show that our algorithm EBACross allows to extract high quality biclusters with highly correlated genes. These biclusters are significant with specific GO term and their genes are particularly involved in specific ontology structure.

To refine the search mechanisms and improve the quality of the extracted biclusters in our future works, we plan to integrate biological knowledge in the research process by benefiting from the help of biologists.

References

1. Ayadi, W., Maatouk, O., Bouziri, H.: Evolutionary biclustering algorithm of gene expression data. In: The Proceedings of the 23th International Workshop on Database and Expert Systems Applications, DEXA 2012, pp. 206–210. IEEE, Vienna (2012)
2. Ben-Dor, A., Chor, B., Karp, R., Yakhini, Z.: Discovering local structure in gene expression data: the order-preserving submatrix problem. In: RECOMB 2002: Proceedings of the Sixth Annual International Conference on Computational Biology, pp. 49–57 (2002)
3. Bergmann, S., Ihmels, J., Barkai, N.: Defining transcription modules using large-scale gene expression data. Bioinformatics 20(13), 1993–2003 (2004)
4. Berrer, D., Dubitzky, W., Draghici, S.: A practical approach to microarray data analysis, ch. 1, pp. 46–53. Kluwer Academic Publishers (2003)
5. Cheng, Y., Church, G.M.: Biclustering of expression data. In: Proceedings of the Eighth International Conference on Intelligent Systems for Molecular Biology, pp. 93–103 (2000)
6. Gene Ontology Consortium. Gene ontology: tool for the unification of biology. Nature Genetics, 25, 25–29 (2000)
7. de Castro, P.A.D., de França, F.O., Ferreira, H.M., Von Zuben, F.J.: Applying biclustering to text mining: an immune-inspired approach. In: de Castro, L.N., Von Zuben, F.J., Knidel, H. (eds.) ICARIS 2007. LNCS, vol. 4628, pp. 83–94. Springer, Heidelberg (2007)
8. Divina, F., Aguilar-Ruiz, J.S.: Biclustering of expression data with evolutionary computation. IEEE Transactions on Knowledge & Data Engineering 18(5), 590–602 (2006)
9. Divina, F., Aguilar-Ruiz, J.S.: A multi-objective approach to discover biclusters in microarray data. In: Proceedings of the 9th Annual Conference on Genetic and Evolutionary Computation, pp. 385–392 (2007)

10. Gasch, A.P., Spellman, P.T., Kao, C.M., Carmel-Harel, O., Eisen, M.B., Storz, G., Botstein, D., Brown, P.O.: Genomic expression programs in the response of yeast cells to environmental changes. Molecular Biology of the Cell 11(12), 4241–4257 (2000)

11. Liu, J., Li, Z., Hu, X., Chen, Y.: Biclustering of microarray data with mospo based on crowding distance. BMC Bioinformatics 10(4), 9 (2009)

12. Liu, X., Wang, L.: Computing the maximum similarity bi-clusters of gene expression data. Bioinformatics 23(1), 50–56 (2007)

13. Madeira, S.C., Teixeira, M.C., Sá-Correia, I., Oliveira, A.L.: Identification of regulatory modules in time series gene expression data using a linear time biclustering algorithm. IEEE/ACM Transactions on Computational Biology and Bioinformatics 7(1), 153–165 (2010)

14. Martinez, R., Pasquier, N., Pasquier, C., Collard, M.: Analyse des groupes de gènes co-exprimés (AGGC): un outil automatique pour l'interprétation des expériences de biopuces. In: SFC 2006 Conference (2006)

15. Mitra, S., Banka, H.: Multi-objective evolutionary biclustering ofgene expression data. Pattern Recognition 39(12), 2464–2477 (2006)

16. Murali, T.M., Kasif, S.: Extracting conserved gene expression motifs from gene expression data. In: Pacific Symposium on Biocomputing, vol. 8, pp. 77–88 (2003)

17. Nepomuceno, J.A., Troncoso, A., Aguilar–Ruiz, J.S.: A hybrid metaheuristic for biclustering based on scatter search and genetic algorithms. In: Kadirkamanathan, V., Sanguinetti, G., Girolami, M., Niranjan, M., Noirel, J. (eds.) PRIB 2009. LNCS (LNBI), vol. 5780, pp. 199–210. Springer, Heidelberg (2009)

18. Nepomuceno, J.A., Troncoso, A., Aguilar-Ruiz, J.S.: Evolutionary metaheuristic for biclustering based on linear correlations among genes. In: SAC 2010, Switzerland, pp. 22–26 (2010)

19. Nepomuceno, J.A., Troncoso, A., Aguilar-Ruiz, J.S.: Biclustering of gene expression data by correlation-based scatter search. BioData Mining 4(3) (2011)

20. Prelic, A., Bleuler, S., Zimmermann, P., Wille, A., Bühlmann, P., Gruissem, W., Hennig, L., Thiele, L., Zitzler, E.: A systematic comparison and evaluation of biclustering methods for gene expression data. Bioinformatics 22, 1122–1129 (2006)

21. Robinson, P.N., Wollstein, A., Bohme, U., Beattie, B.: Ontologizing geneexpression microarray data: characterizing clusters with gene ontology. Bioinformatics 20(6), 979–981 (2004)

22. Seridi, K., Jourdan, L., Talbi, G.: Multi-objective evolutionary algorithm for bi-clustering in microarrays data. In: IEEE Congress of Evolutionary Computation, pp. 2593–2599 (2011)

23. Shyama, D., Sumam, M.I.: Application of greedy randomized adaptive search procedure to the biclustering of gene expression data. International Journal of Computer Applications 2(3), 0975–8887 (2010)

24. Tanay, A., Sharan, R., Shamir, R.: Discovering statistically significant biclusters in gene expression data. Bioinformatics 18, 136–144 (2002)

25. Tavazoie, S., Hughes, J.D., Campbell, M.J., Cho, R.J., Church, G.M.: Systematic determination of genetic network architecture. Nature Genetics 22, 281–285 (1999)

26. Valente-Freitas, A., Ayadi, W., Elloumi, M., Oliveira, J.L., Hao, J.K.: A survey on biclustering of gene expression data. In: Biological Knowledge Discovery Handbook: Preprocessing, Mining, and Postprocessing of Biological Data, pp. 591–608 (2013)

SFFS-SW: A Feature Selection Algorithm Exploring the Small-World Properties of GNs

Fábio Fernandes da Rocha Vicente[1,2] and Fabrício Martins Lopes[1]

[1] Federal University of Technology, Paraná, Brazil
[2] Institute of Mathematics and Statistics,University of São Paulo, São Paulo, Brazil
{fabiofernandes,fabricio}@utfpr.edu.br

Abstract. In recent years, several methods for gene networks (GNs) inference from expression data have been developed. Also, models of data integration (as protein-protein and protein-DNA) are nowadays broadly used to face the problem of few amount of expression data. Moreover, it is well known that biological networks conserve some topological properties. The small-world topology is a common arrangement in nature found both in biological and non-biological phenomena. However, in general this information is not used by GNs inference methods. In this work we proposed a new GNs inference algorithm that combines topological features and expression data. The algorithm outperforms the approach that uses only expression data both in accuracy and measures of recovered network.

Keywords: small-world, gene networks, feature selection, graph theory, pattern recognition, bioinformatics.

1 Introduction

Complex networks systems is a very common phenomenon. In fact, we live in a universe of things that can be seen as complex networks [1–3]. These real systems are both biological and non-biological such as Internet, social interactions, physical systems, infection dynamics, regulatory networks, to cite but a few [3–7]. The occurrence of certain *specific topologies* has been observed and characterized in several research fields. These works points that diverse networks of natural phenomenon are not random but follow some particular arrangements [2–4, 8–10]. Thus, it is necessary to describe these distinct *specific topologies* in some manner to better understand the differences between them. In this way two aspects are important: *characterization* and *representation* [11]. A network can be characterized through a feature vector composed by network measurements, such as average vertex degree, average path length, degree distribution, etc., in which the network is said *mapped* to the feature vector. Thus, the feature vector can be used to group the several networks topologies into classes. On other hand, the inverse way is commonly impossible and the original network cannot be recovered from the feature vector. However, when the network can be recovered the mapping is said to provide a *representation*. Examples of representation are the adjacency list and the adjacency matrix[11, 12].

M. Comin et al. (Eds.): PRIB 2014, LNBI 8626, pp. 60–71, 2014.

The study of complex networks can occur into two scenarios: In the first, the topology is determined by construction through a theoretical model thus, both the complete network and its properties can be clearly known a priori. The second group is composed by those networks for which the underlying construction rules (if one exists) are unknown and even the graph can be partially observed. In this case the topology is described in terms of its features. Frequently, both the rules of construction and the complete network are unknown as for instance, real neuronal network, gene networks (GNs) and social networks. The exception can occurs on planed networks as artificial neuronal networks, subway, power distribution, computer networks, etc.

When matter is life, the class of topology is commonly unknown. However, biological networks conserve properties between organisms and it is not very distinct of non-biological phenomenon [13]. Therefore, the research on this area try to determine some features that can *characterize* biological networks sometimes adopting some construction model as reference. Distinct descriptions has been proposed to biological networks as scale-free [2] for example in *yeast* [13] and metabolic networks [13]. Also a hierarchic structure was observed in *E.coli* [14]. In special, the small-world (SW) topology [15, 16] has been observed in several areas and seems to be a common phenomenon both in physical events, social and biological networks. Some examples are: seismic events [7], subway station distribution [5], social networks [15, 17], value dynamics in financial market [18], epidemics [19], neuronal networks [4] and brain networks [6].

In particular, it was adopted in this work the SW construction model proposed by Watts and Strogatz [20, 16]. This model has two main properties: small average path length and high average clustering coefficient. These SW features has been observed on biological networks. For instance human protein-protein interaction network (PPI) presented SW properties [21] and a study of networks of 43 organisms presented a higher clustering coefficient [22].

The research of gene networks is broadly used to better understand living organisms. Moreover, since many gene interactions remains unknown, the inference of gene networks from expression data has been used to discover new interactions [23–27]. Moreover, the topology of living organisms should conserve some structure, following some particular features. Considering that the inference of gene networks is an inverse problem where more than one network could produce equivalent data [28, 26], the use of topology could help the search algorithms by avoiding improbable biological structures. An algorithm guided by scale-free (Barabási-Albert [2]) topology achieved better accuracy on inference [29]. Thus, the topology can be an important component for GNs inference. This work presents a new algorithm for GNs inference whose criterion function is based both in expression data and topological features of SW networks.

Methodology

The real structure of gene networks (GNs) is commonly unknown and the observation of physical relationships between components (protein-DNA, protein-protein, etc.) is expensive and some times hard to obtain. Therefore, the network is frequently inferred from expression. The reasoning behind this approaches is that the set of relationships between cellular components (i.e. the network) produces the output observed expression. The challenge is to recover the network from the observed data [23, 26]. It was adopted in this work the Probabilistic Boolean Networks (PBN) model [23] to represent the gene network. PBN is a probabilistic approach derived from Boolean Networks (BN) model, first introduced by Kauffman [30]. The genes of a BN are represented by variables and can just assume discrete values, typically 1 (the gene is up regulated) and 0 (the gene is down regulated). In other words, the value correspond to the expression of that gene in a given observation t. The value of one given variable in the sample t is named *state of the variable* and the set of values of all variables in the sample t is called *state of the system*. The relationship between genes is represented by boolean functions. If the samples is given through a time series, the set of boolean functions applied to the system state at observation t determines the state of the system in the next observation $t + 1$. Thus, the transitions between states can be deterministic (BN) if there is no changes in the set of boolean functions or probabilistic (PBN) where each variable is associated to set of boolean functions with a probability of choice at each observation.

In this work we address the problem of selecting a set of *predictor* genes (i.e. features) in the sample t that can better be used to classify the state value of a given *target* gene on a sample $t+1$ based on the state of the set of predictors in the time sample t. In this context, a naive algorithm could perform an evaluation over all possible combinations of $n - 1$ genes taken k at time, with $k = 1, 2, \ldots, n-1$, leading to an exponential increasing of the search space with the increment of n. Ir order to avoid this computational complexity we have adopted a feature selection method.

Feature Selection Algorithm

Regarding the classification task, a feature selection approach try to select a subset of features that produce the better classification of the observed classes. In this way, a feature selection algorithm requires two components: a *criterion function* that assign a value to a subset of features and a *search algorithm* whose objective is performed in order to find a subset of features that minimize/maximize the criterion function. The SFFS algorithm [31] and other alternative versions to the canonical SFFS has been applied on GNs inference [32, 33]. Thus, we adopted the SFFS search strategy on this work.

Criterion Function

We defined a criterion function that includes both the gene expression data and a topological features. The expression values are used to compute the Mean Conditional Entropy (MCE) which is presented below.

Mean Conditional Entropy. In some problems it is necessary to assign a measure of uncertainty to a given random variable Y after the observation of another random variable X. For instance, the uncertainty about the value of a target gene Y after the observation of the value of another gene X. The conditional entropy of Y given x is defined as:

$$H(Y|x) = -\sum_{y \in Y} P(y|x)log[P(y|x)] \tag{1}$$

The Mean Conditional Entropy [32] is the weighted average of $H(Y|x)$ for each $x \in X$.

$$H(Y|X) = \sum_{x \in X} H(Y|x)P(x) \tag{2}$$

In the common context of GNs inference, the algorithm must search for those variables (i.e. features) that minimizes the MCE even if using few observations. Furthermore, even one had large observation set, it could be expected that some system states does not occur because they can be rare [26, 34, 23]. Thus, the number of *observed* instances (N) is commonly much lower then the number of *possible* instances (M). Thereby, to face this problem the MCE is computed with penalization of the non-observed instances as defined in [32]:

$$H(Y|X) = \frac{\alpha(M-N)H(Y) + \sum_{i=1}^{N}(f_i + \alpha)H(Y|X = x_i)}{\alpha M + T} \tag{3}$$

where T is the number of samples, f_i the absolute frequency of x_i and α is a parameter to determine the weight of the penalization.

Small-World Networks. The *small-world* network proposed by Watts and Strogatz [20] allows to model the increasing of randomness in a regular network. The SW construction proposed by Watts and Strogatz is defined as follows: First set N as the number of vertices and k as the average degree. Start with a one-dimensional ring lattice with each vertex connected to $2k$ neighbors. Then, for each vertex rewire each edge with probability p. Thus, p defines the global randomness of the network which ranges from $p = 0$ (regular original lattice) to $p = 1$ (totally random).

The two main features of this networks are the *clustering coefficient* (C) and *path length* (L). The clustering coefficient (Eq. 4) of a vertex C_v is defined as follows: given a vertex v, select its m neighbors. Then, count the number of edges η_v between the m neighbors of v (i.e. except the edges with v). Let $\eta_m = m(m-1)/2$ denote the maximum number of edges in the sub-graph with m vertices. Then, compute $C_v = \eta_v/\eta_m$. Thus, the network clustering coefficient is given by averaging over all vertices:

$$C = \frac{1}{N} \sum_{v \in V} C_v \tag{4}$$

The path length L_{ij} is defined as the number edges in the *shortest path* between two vertices v_i and v_j. The network path length L is the average over all pair of vertices. Let λ denote the number of paths in the network.

$$L = \frac{1}{\lambda} \sum_{v_i,v_j \in V} L_{ij} \qquad (5)$$

MCE-SW Criterion Function. We defined a criterion function based on the MCE and the topological properties of SW networks (MCE-SW). The MCE-SW is a linear combination of MCE and SW features. The topological part is composed by the two SW features described by Watts and Strogatz [16].

The clustering coefficient is in the interval $[0,1]$ as defined on Equation 4. The path length L is a value greater than 1 without upper limit since N is a network parameter. The theoretical maximum path length for a given pair v_i, v_j is $N-1$ in a network of size N. However, the path length in a SW network is frequently very lower than that maximum. Thus, L is normalized through *max-min* normalization (Eq. 6). Where $min = 1$ and max was estimated by sampling 1000 SW networks with the same k,N and p and taking the maximum path length. Also, if eventually L is greater than max than we set $L = 1$.

$$Normalized(L) = \frac{L - min}{max - min} \qquad (6)$$

Form this point we will refer to the $Normalized(L)$ simply as L. The MCE-SW criterion function has one parameter $w \in [0,1]$ which is the *weight* of topological features. We set half of w to each topological measure $w_1 = w/2$ and set weight of MCE $w_2 = 1 - w$. Thus, the MCE-SW criterion function of two variables x and y is defined as:

$$\begin{aligned} \text{MCE-SW}_{y,x} &= w_2 \times MCE_{x,y} + w_1 \times L - w_1 \times C \\ &= w_2 \times MCE_{x,y} + w_1 \times (L - C) \end{aligned} \qquad (7)$$

Where $C \in [0,1]$ and $L \in [0,1]$. Since, $L - C \in [-1,1]$ we rescale $(L-C)$ in $[0,1]$ to maintain a positive score.

$$SW = \frac{(L - C) + 1}{2} \qquad (8)$$

Thus, MCE-SW became:

$$\text{MCE-SW}_{y,x} = w_2 \times MCE_{x,y} + w_1 \times SW \qquad (9)$$

SFFS-SW Feature Selection Algorithm

The proposed search algorithm is composed by two distinct steps. At the fist step the network is inferred from expression data by using MCE as criterion function, which guarantees an initial topology. On the second step each vertex is revisited at the same order they were visited in the first phase and the MCE-SW criterion function is applied. As can be seen in Equation 9 the part MCE gives the *local* contribution of the edges between a gene y and its predictors. In other words, MCE does not take into account how the the global topology if affected by those edges. Differently, the topological part measures how the local changes influences the global properties of the network. Thus, the second phase of the algorithm readdress the inferred edges in terms of both local and global topological effects.

The Algorithm 1 receive N genes G, the expression data set D, the maximum number of features mf the SFFS will search for each target Y and the weight of topological measurements w. Once the set or predictors of a target gene Y is defined by SFFS-SW algorithm, the edges from predictors to target are added and thus, the network is updated. At each iteration of the second step

---

input : G,D,mf,w
output: network
var list: predictor-set[N]
First step
foreach *target* $Y \in G$ **do**
    predictor-set[Y] $\leftarrow$ SFFS-MCE(Y, G, D, mf)
    network $\leftarrow$ update-network(predictor-set[Y])
end
Second step
foreach *target* $Y \in G$ **do**
    predictor-set[Y] $\leftarrow$ SFFS-SW(Y, G, D, mf, w)
    network $\leftarrow$ update-network(predictor-set[Y])
end

Algorithm 1. Main function of the inference algorithm

of Algorithm 1, the SFFS method is performed to search for the best subset of size $q \leq mf$. The edges from a feature set X to the target gene Y are temporarily added to the graph in order to compute the *topological features* and the *criterion function*. After the criterion function is computed the edges are removed. At the end, when the best subset is finally chosen the edges are permanently added in the network. The Algorithm 2 shows the computation of the criterion function value for a given feature set X. The algorithm starts by removing all previously inferred predictor of that target variable in the first step of Algorithm 1. The inclusion of each feature on this execution take into account how much the added characteristic affects both the MCE and the network topology by considering all other previously inferred edges. In other words, the predictor set of each target is re-inferred from an empty set, based on the possible changes in topology.

```
input   : Y,X ∈ G, D, w
output: criterion function value
MCE ← H(Y|X)
network ← add-edges(X, Y)
Compute L, C and SW
MCE-SW_{X,Y} = w_2 × MCE_{x,y} + w_1 × SW
network ← remove-edges(X, Y)
return MCE-SW_{X,Y}
```

Algorithm 2. MCE-SW criterion function computation

Computational Complexity. Assuming a network with N vertex and maximum degree k, the complexity of SFFS algorithm is performed N times (one for each target), each with cost $O(2^k)$. The complexity of computation of the average clustering coefficient part depends on the number of neighbors k_v of each vertex v. The algorithm must to sum the number of edges between the k_v neighbors of v. Thus, the maximum number of operation for a given vertex is k^2, The shortest path between two vertices is computed in $O(N^2)$ and the average shortest path is computed to N vertex in $O(N^3)$. Thus, for each subset of predictors (or for each target) the algorithm calculates the average clustering coefficient and average path length procedures performing $k^2 + N^3$ operations. Thus, the SFFS performs $2^k × (k^2 + N^3)$ operations. Since in the context of GNs inference k is limited to a small value the total cost of one run of SFFS algorithm is $O(N^3)$. The SFFS algorithm is executed N times (one for each target) on each step of algorithm. Thus the total complexity of algorithm is $O(N^4)$. The computational cost is justified by the increasing in the performance on recovering a network with better topological features.

Validation. In order to evaluate the method we used the Artificial Gene Networks (AGN) simulation and validation model [35, 36] to generate both the networks and the expression data. The AGN was used to create Watts and Strogatz SW networks and to simulate the corresponding output signal by considering a probabilistic boolean network model. To evaluate the performance of the proposed algorithm we computed the following confusion matrix:

Table 1. Confusion matrix. TP = true positive, FN = false negative, FP = false positive, TN = true negative.

Edge	Inferred	Not Inferred
Present	TP	FN
Absent	FP	TN

Commonly, in GNs inference it is preferable to predict a low number of edges with high precision than a high number of edges with low precision. This has a practical aspect: the inferred relationships will probably be validated through some expensive biochemical experiment. Thus, the set of predicted edges can be as a set of hypothesis to be tested. However, an ideal algorithm model should recover edges with high precision without missing existing edges. To evaluate these properties we compute the PPV (Positive Predictive Value, also know as *precision*) and Sensitivity (also know as *recall* or True Positive Rate – TPR). We adopted also the *similarity* measure which is a geometrical average of precision and recall.

$$PPV = \frac{TP}{(TP + FP)}$$

$$Sensitivity = TPR = \frac{TP}{(TP + FN)} \tag{10}$$

$$Similarity(A, B) = \sqrt{PPV \cdot Sensitivity}$$

Since we are mostly interested on the inference of networks that follow small-world properties, we also adopted the computation of trajectories for both *clustering coefficient* and *path length* at each step of the inference algorithm.

Results and Discussion

This section presents the experimental results of the proposed methodology when applied to SW networks. We evaluate our methods in networks with 100 vertices and average edges k varying from 1 up to 4. We set the rewiring probability to $p = 0.01$ and the threshold of the criterion function to 0.3. We performed 10 executions to each configuration. On each execution produced a network in the first phase (*recovered-1*). Then, the second step of the proposed algorithm was executed with weights $0.2, 0.4, 0.6, 0.8, 1.0$. For each weight the search always starts from the same *recovered-1* network of that configuration.

It is important that the topological features do not suppress the expression information but on the contrary, be well combined in order to increase the accuracy and the networks properties. We find a small variation in accuracy as the topological information weight vary from $0.2 - 0.6$ (Figure 1).

The algorithm frequently achieves the highest precision values when $w - 0.8$ (see Table 2). There is a gain in precision given by the inclusion of topological information for any value of k. It is interesting to note that when the weight is 1.0 the PPV decreases fast showing that both expression and topological information are essential to inference. There is also an improvement in sensitivity when SW features is used. The exception occurs in the highest connected network (k=4) were the value is not altered. Thus, this also suggests an interdependence between topology and the expression. Moreover, the topological features does not suppress the expression information, but in contrary, the search is biased through the combination of these two elements. the inference is not dominated by topological features it is just *biased* by the new criterion function.

We also analyzed the trajectory of clustering coefficient (C) and path length (L) – Figure 2. It could be observed distinct behaviors between C and L and also an improvement on the small-world features of the recovered network.

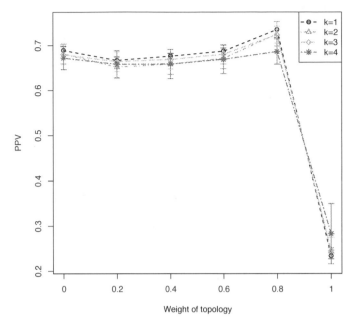

Fig. 1. Positive Predictive Value (PPV). The algorithm reaches the highest value with w=0.8 for any K. The precision decreases fast when the algorithm does not consider the MCE (w=1.0) showing that both expression and topology is necessary to increase precision.

Table 2. Precision (PPV), Sensitivity (Recall) and Similarity for $k = 1, 2, 3, 4$ and weight$=0, 0.2, 0.4, 0.6, 0.8, 1.0$

		weight					
		0	0.2	0.4	0.6	0.8	1.0
k=1	PPV	0.56	0.54	0.55	0.56	**0.60**	0.10
	Sensitivity	0.78	**0.80**	**0.80**	**0.80**	**0.80**	0.09
	Similarity	0.76	0.75	0.76	0.76	**0.78**	0.20
k=2	PPV	0.60	0.57	0.58	0.59	**0.65**	0.11
	Sensitivity	0.58	**0.59**	0.58	0.58	0.58	0.05
	Similarity	0.70	0.69	0.69	0.70	**0.72**	0.18
k=3	PPV	0.62	0.60	0.61	0.62	0.67	0.13
	Sensitivity	0.44	**0.45**	**0.45**	0.44	0.44	0.10
	Similarity	0.65	0.64	0.64	0.65	**0.66**	0.10
k=4	PPV	0.67	0.66	0.66	0.67	**0.69**	0.28
	Sensitivity	**0.39**	**0.39**	**0.39**	**0.39**	0.35	0.01
	Similarity	0.64	0.64	0.63	**0.64**	0.62	0.12

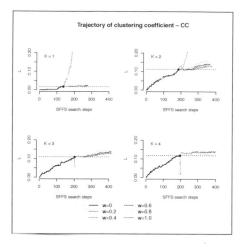
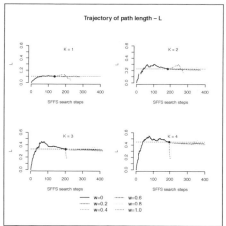

Fig. 2. Left: trajectory of the average clustering coefficient. The first step of the algorithm (SFFS-MCE) is printed in black. The achieved value of the SFFS-SW step is higher than the final value of the SFFS-MCE step (exceptions: *weight* = 1.0 with $k = 3$ and $k = 4$). **Right:** trajectory of the average path length (normalized in the range [0,1]). The SFFS-SW achieves a lower value than the SFFS-MCE. The exception is *weight* = 0.8.

With respect to *clustering coefficient* C the value starts from zero, gradually increases until it achieves a point. In the second step the value increases when SFFS-SW is applied, showing the relevance of this topological feature.

With respect to *path length*, the algorithm presented a distinct behavior. Like C, on the first phase L starts from zero. However, it can be observed that L increases fast at the first steps then, it decreases and converges to an smallest path length. This shows that the SFFS-MCE algorithm drives the search to a small path length. Following this reasoning, the bias given by the new criterion function drives the path length to a lower value. However, the trajectory maintains almost stable in respect to path length.

The MCE-SW criterion function improve the features of the inferred network by increasing the clustering coefficient and maintaining the achieved path length. The algorithm reach different values as average degree increases.

Conclusion

In this work we presented a new feature selection method, called SFFS-SW for the inference of GNs. The algorithm uses both expression data and two topological features of SW networks: high average clustering coefficient and low average path length. The SFFS-SW combines the Mean Conditional Entropy (MCE), network clustering coefficient and the network path length through a criterion function. The search is performed through an SFFS algorithm in two steps: The first step uses only expression to recover a network. On the second step the network is inferred by considering both MCE and SW features. The proposed algorithm inferred networks with higher precision and recall than the strategy that uses only expression data. Moreover, the inferred networks are biased to topology with higher clustering coefficient and lower average path length than networks inferred without topological information. Thus, even when precision and recall are very similar between topological and non-topological approaches, the network features are biased to SW topology in SFFS-SW. The reasoning is that it is important not only to find a set of correct edges and discard incorrect ones, but to find those correct edges that are also consistent to the network topology. Finally, the results indicate that topological information is important both on the inference process and evaluation of results.

Acknowledgement. We would like to thanks the financial support of FAPESP grant 2011/50761-2, CNPq, CAPES, NAP eScience - PRP - USP, and Fundação Araucária.

References

1. Latora, V., Marchiori, M.: Efficient Behavior of Small-World Networks. Physical Review Letters 87(19), 198701 (2001)
2. Albert, R., Barabási, A.L.: Statistical mechanics of complex networks. Reviews of Modern Physics 74(1), 47–97 (2002)
3. Easley, D., Kleinberg, J.: Networks, Crowds, and Markets: Reasoning about a Highly Connected World. Cambridge University Press (2012)
4. Lago-Fernández, et al.: Fast response and temporal coherent oscillations in small-world networks. Physical Review Letters 84(12), 2758—2761 (March 2000)
5. Latora, V., Marchiori, M.: Is the Boston subway a small-world network? Physica A: Statistical Mechanics and its Applications 314(1-4), 109–113 (2002)
6. Bassett, D.S., Bullmore, E.: Small-world brain networks. The Neuroscientist: a Review Journal Bringing Neurobiology, Neurology and Psychiatry 12(6), 512–523 (2006)
7. Baek, W.H., et al.: Analysis of topological properties in a seismic network. Physica A: Statistical Mechanics and its Applications 391(6), 2279–2285 (2012)
8. Amaral, L., et al.: Classes of small-world networks. Proceedings of the National Academy of Sciences of the United States of America 97(21), 11149–11152 (2000)
9. Vázquez, A., et al.: The topological relationship between the large-scale attributes and local interaction patterns of complex networks. Proceedings of the National Academy of Sciences of the United States of America 101(52), 17940–17945 (2004)
10. Brockmann, D., Helbing, D.: The Hidden Geometry of Complex, Network-Driven Contagion Phenomena. Science 342(6164), 1337–1342 (2013)
11. da Costa, L.F., et al.: Characterization of complex networks: a survey of measurements. Advances in Physics 56(1), 167–242 (2007)
12. Pavlopoulos, G., et al.: Using graph theory to analyze biological networks. BioData Mining 4(1), 10 (2011)
13. Jeong, H., et al.: The large-scale organization of metabolic networks. Nature 407(6804), 651–654 (2000)
14. Ma, H.W., et al.: Hierarchical structure and modules in the Escherichia coli transcriptional regulatory network revealed by a new top-down approach. BMC Bioinformatics 5, 199 (2004)

15. Milgram, S.: The Small-World Problem. Psychology Today 1(1), 61–67 (1967)
16. Strogatz, S.H.: Exploring complex networks. Nature 410(6825), 268–276 (2001)
17. Newman, M.E., Watts, D.J.: Scaling and percolation in the small-world network model. Physical review. Physical review. E, Statistical Physics, Plasmas, Fluids, and Related Interdisciplinary Topics 60(6 pt. B), 7332–7342 (1999)
18. Zhao, H., et al.: Self-organizing Ising model of artificial financial markets with small-world network topology. EPL (Europhysics Letters) 101(1), 18001 (2013)
19. Moore, C., Newman, M.E.: Epidemics and percolation in small-world networks. Physical review. E, Statistical Physics, Plasmas, Fluids, and Related Interdisciplinary Topics 61(5 pt. B), 5678–5682 (2000)
20. Watts, D.J., Strogatz, S.H.: Collective dynamics of 'small-world' networks. Nature 393(6684), 440–442 (1998)
21. Assenov, Y., et al.: Computing topological parameters of biological networks. Bioinformatics 24(2), 282–284 (2008)
22. Ravasz, E., et al.: Hierarchical organization of modularity in metabolic networks. Science 297(5586), 1551–1555 (2002)
23. Shmulevich, I., Dougherty, E.: From Boolean to probabilistic Boolean networks as models of genetic regulatory networks. Proceedings of the IEEE 90(11), 1778–1792 (2002)
24. Karlebach, G., Shamir, R.: Modelling and analysis of gene regulatory networks. Nature Reviews. Molecular Cell Biology 9(10), 770–780 (2008)
25. Baralla, A., Mentzen, W.I., de la Fuente, A.: Inferring gene networks: dream or nightmare? Annals of the New York Academy of Sciences 1158, 246–256 (2009)
26. Dougherty, E.R., Bittner, M.L.: Causality, randomness, intelligibility, and the epistemology of the cell. Current Genomics 11(4), 221–237 (2010)
27. Lopes, F.M., Ray, S.S., Hashimoto, R.F., Cesar Jr., R.M.: Entropic biological score: a cell cycle investigation for GRNs inference. Gene 541(2), 129–137 (2014)
28. Dougherty, E.R.: Validation of inference procedures for gene regulatory networks. Current Genomics 8(6), 351–359 (2007)
29. Lopes, F.M., Martins Jr., D.C., Barrera, J., Cesar Jr., R.M.: A feature selection technique for inference of graphs from their known topological properties: Revealing scale-free gene regulatory networks. Information Sciences 272, 1–15 (2014)
30. Kauffman, S.A.: Metabolic stability and epigenesis in randomly constructed genetic nets. Journal of Theoretical Biology 22(3), 437–467 (1969)
31. Pudil, P., Novovičová, J., Kittler, J.: Floating search methods in feature-selection. Pattern Recognition Letters 15(11), 1119–1125 (1994)
32. Lopes, F.M., Martins Jr., D.C., Cesar-Jr, R.M.: Feature selection environment for genomic applications. BMC Bioinformatics 9(1), 451 (2008)
33. Lopes, F.M., Martins Jr., D.C., Barrera, J., Cesar Jr., R.M.: SFFS-MR: a floating search strategy for GRNs inference. In: Dijkstra, T.M.H., Tsivtsivadze, E., Marchiori, E., Heskes, T. (eds.) PRIB 2010. LNCS, vol. 6282, pp. 407–418. Springer, Heidelberg (2010)
34. Yu, L., Watterson, S., Marshall, S., Ghazal, P.: Inferring Boolean networks with perturbation from sparse gene expression data: a general model applied to the interferon regulatory network. Molecular BioSystems 4(10), 1024–1030 (2008)
35. Lopes, F.M., Cesar Jr., R.M., da Costa, L.F.: AGN simulation and validation model. In: Bazzan, A.L.C., Craven, M., Martins, N.F. (eds.) BSB 2008. LNCS (LNBI), vol. 5167, pp. 169–173. Springer, Heidelberg (2008)
36. Lopes, F.M., Cesar Jr., R.M., da Costa, L.F.: Gene expression complex networks: Synthesis, identification, and analysis. Journal of Computational Biology 18(10), 1353–1367 (2011)

CytomicsDB: A Metadata-Based Storage and Retrieval Approach for High-Throughput Screening Experiments

E. Larios[1], Y. Zhang[2], L. Cao[1], and F.J. Verbeek[1]

[1] Section Imaging and Bioinformatics, LIACS, Leiden University, Leiden, The Netherlands
{e.larios.vargas,l.cao,f.j.verbeek}@liacs.leidenuniv.nl
[2] Centrum Wiskunde & Informatica, Amsterdam, The Netherlands
{Ying.Zhang}@cwi.nl

Abstract. In Cytomics, the study of cellular systems at the single cell level, High-Throughput Screening (HTS) techniques have been developed to implement the testing of hundreds to thousands of conditions applied to several or up to millions of cells in a single experiment.

Recent technological developments of imaging systems and robotics have lead to an exponential increase in data volumes generated in HTS-experiments. This is pushing forward the need for a semantically oriented bioinformatics approach capable of storing large volume of linked metadata, handling a diversity of data formats, and querying data in order to extract meaning from the experiments performed.

This paper describes our research in developing CytomicsDB, a modern RDBMS based platform, designed to provide an architecture capable of dealing with the computational requirements involved in high-throughput content. CytomicsDB supports web services and collaborative infrastructure in order to perform further exploration of linked information generated in each experiment.

The objective of this system is to build a semantic layer over the data so as to enable querying metadata and at the same time allowing scientists to integrate new tools and APIs taking care of the image and data analysis. The results will become part of the metadata of the whole HTS experiment and will be available for semantic post analysis.

1 Introduction

High-Throughput Screening (HTS) is a well-established process in drug discovery for pharma and biotechnology companies and is now also being set up for basic and applied research in academia and some research hospitals [10]. Recent developments in microscopy systems and robotics enabled large-scale screening of cellular systems. A popular screen setup is automated time-lapse confocal image acquisition which enables capturing of e.g. high content subcellular information (derived as features) or dynamic aspects like cell migration. Cells are exposed to hundreds and even thousands of different conditions using one or several multiwell (96, 384, 1536) plates. This typically results in 20-40 GB of data consisting of in the order of 100,000 - 200,000 images in an overnight experiment.

In cytometry, HTS-experiments are usually employed in the context of functional analysis, closing the gap between genomics-proteomics and functional responses on the

M. Comin et al. (Eds.): PRIB 2014, LNBI 8626, pp. 72–84, 2014.

cellular level. Examples are genome wide siRNA screens, where all existing genes are lowered in activity one at a time using siRNA mediated knock down followed by some cellular-level phenotypic readout, e.g., cell migration speed, focal adhesion dynamics, subcellular morphological changes, cell death.

A next step in the HTS-experiment pipeline is image quantification using image analysis software tools. In this manner, biological hypothesis can be statistically tested using the quantification results from the image analysis stage, and can depict an objective understanding of the cell response to various treatments or exposures.

In a typical HTS workflow, spreadsheet applications are commonly used for book-keeping all information related to the design of the multiwell-imaging plates, image analysis quantification results and even statistical analysis results. This approach has many drawbacks. Firstly, it is extremely difficult to link the data produced during the different stages of an HTS experiments, such as linking the images generated in the HTS experiment and the metadata collected during the design of the plate layout. Secondly, it is highly prone to man made errors. The lack of standards, formats and a centralized place for storing the information makes it difficult to promote a collaborative environment with or between research groups. Finally, spreadsheet applications are not suitable for knowledge discovery, as they do not allow to combine sophisticated visualization and querying of the (meta)data previously stored.

In our previous work [8], we presented the initial design of a platform for managing and analyzing HTS images resulting from cytomics screens taking the automated HTS workflow as a starting point. This platform *seamlessly* integrates the whole HTS workflow into a single system. The platform relies on a modern relational database system to store user data and process user requests, while providing a convenient web interface to end-users. Using this platform, the overall workload of HTS experiments, from experiment design to data analysis, can be significantly reduced. Additionally, the platform provides the potential for data integration to accomplish genotype-to-phenotype modeling studies. In this paper, the initial design, particularly, the database model, has been rigorously revised and generalised to manage all kinds of metadata produced by automated HTS systems. We call our system *CytomicsDB*, which is designed as a user oriented platform but considers the HTS workflow as a template for managing, visualizing and querying the metadata.

Current software and architectures for HTS are mostly based on generic Lab Information Management Systems (LIMS) [12], which face significant challenges to accessing, analyzing, and sharing the data required to drive day-to-day processes within the laboratory. Furthermore, the limited connectivity to other legacy systems and poor visualization of the data is an obstacle to extract new insights from the data stored, and cause a deep impact in the efficiency of the HTS experiment. Comparing with the existing LIMS systems, CytomicsDB has a number of important advantages:

1. Ease of promoting scientific collaborations. Since all data in CytomicsDB are centralized, granting access to collaborators or sharing information has been made simple;
2. Flexibility for integration with other legacy systems. It it common to use external APIs for performing image and data analysis results, such as Weka, PRTools. In the

design of the architecture of CytomicsDB, special care has been taken to assure the possibility of invoking external API through web services.

3. The web-based architecture allows its users to easily access to their experiments data from wherever and at any time. The architecture also allows the whole or parts of the system to be smoothly moved to a Cloud based environment.
4. The capability to drill-down through experiments' metadata due to the metadata-based approach.
5. A single interface for visualization of all experiments data, include raw images, metadata and analysis results.
6. Pattern recognition (PR) within an experiment and PR across HTS experiments.

To sum up, the contributions of this work include:

1. Metadata organization in an HTS experiment (Section 2).
2. Data modeling and storage (Section 3).
3. A case study in endocytosis of EGFR, describing how a Metadata-based RDBMS approach can facilitate the identification of EGFR dynamics and classification of EGFR phenotype stages (Section 4).

Finally, in Section 5 we discuss related work and in Section 6 present our conclusions.

2 Metadata Organization in an HTS Experiment

The metadata of an HTP experiment consists of a variety of types and formats and has been grouped in five levels as showed in Figure 1: Project, Experiment, Plate - Wells, Video/Images and Measurements. These levels contain each other in a cascade fashion, for instance: [1] Project contains [1..n] Experiments, [1] Experiment contains [1..n] Plates, [1] Plate contains [24,48,96,384] Wells, [1] Well contains [1..n] Video/Images and finally [1] Well contains [1..n] measurements.

Project. This level contains a title which describes the aim of the project, the duration, the author, etc. When a project is created, its creator becomes its administrator and is possible to grant access to another scientist in order to promote a collaborative environment.

Experiment. Figure 2 shows the structure of the metadata contained in the Experiment level. This level is divided in Hardware and Type of Experiment. Firstly, the metadata associated to the hardware correspond to the microscope and the imaging technique used. Depending on which microscope is used, the set of imaging techniques differs. For instance, the imaging techniques available for a Becton Dickinson (BD) Pathway microscope are EPI, Spinning disk or Bright Field, but in a Nikon TE 2000-e microscope it is possible to use: FRAP, FRET, EPI, Confocal, Spectral or DIC. Secondly, the metadata associated to the type of experiment can be separated in four groups: (1) Fixed or Live experiment including a 2D or 3D option for each case; (2) Assay type, in this case there are the following options: migration/invasion, proliferation, primary tumor, apoptosis and sub cellular perturbations; (3) Species, the options available are: human, rat, mouse and zebrafish; and (4) Cell / Tissue origin, considering in this area: primary, cell line, iPSC, stem cel, biopsy, etc.

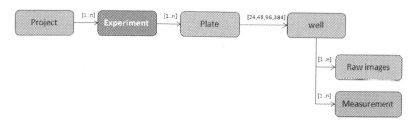

Fig. 1. Structure of the metadata in an HTS experiment

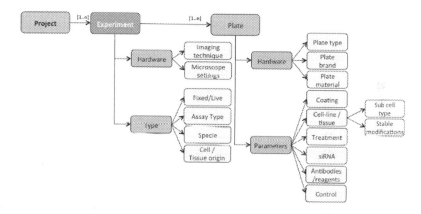

Fig. 2. Structure of the Experiment and Plate metadata

Plate-Wells. This level is also divided into two groups: metadata about the Hardware and about the Parameters used. Figure 2 shows an UML diagram of the structure of these two groups. The Hardware sub level includes information about the plate type, the brand and the fabrication material. The level of Parameters includes information about (1) Coating, (2) Cell-line / tissue, (3) treatment, (4) siRNA, (5) Antibodies / reagents and (6) Parameters of control or comments. The metadata of Wells is a subset of metadata of the plate level. For instance, in a 8x12 wells plate, different wells can have a subset of the parameters assigned to the whole plate. This level is also associated with the output of the HTS process (Raw Video / Images) and with the results of the image and data analysis phase which is also called measurements.

Part of the metadata at this level is critical information that should be verified and validated when it is uploaded. For instance, The parental cell line/tissue, or the treatment and its concentration are just two cases which the entry is verified in a first instance (obligatory data) and then they are validated with the information pre loaded in the imaging database. In order to keep the consistency of the metadata it is necessary to validate each entry and when a new value is detected the administrator of the platform is in charge of accepting this new entry as valid or correct to the right value if it is necessary. The consistency in the metadata is a key task in the imaging database because the obligatory data will be further used as a controlled vocabulary for querying.

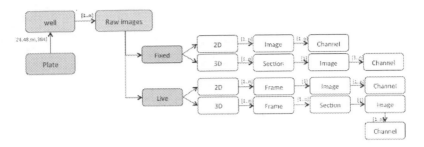

Fig. 3. Structure of the Raw Images metadata

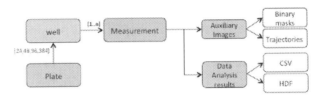

Fig. 4. Structure of the Measurements metadata

Raw Images. Raw images are obtained after image acquisition with automated microscopy systems. These images are the basis for the image analysis which results in quantitative data used for hypothesis testing. The response of the cells is recorded through time-lapse microscopy imaging and the resulting image sequences are the basis for the image analysis. The structure of an image file depends on the type of experiment (Fixed/Live) and the microscopy technique used in the experiment. Currently, four types of structures are supported (cf. Figure 3) [8]:

1. 2D (XY): this structure corresponds to one frame containing one image which is composed of multiple channels ([1]Frame - [1]Image - [1..n]Channels).
2. 2D+T (XY+T): this structure corresponds to one video with multiple frames. Each frame contains one image composed of multiple channels ([1]Video - [1..n]Frame - [1]Image - [1..n]Channels).
3. 3D (XYZ): this structure corresponds to one frame with multiple sections. Each section contains one image composed of multiple channels ([1]Frame - [1..n] Sections - [1]Image - [1..n]Channels).
4. 3D+T(XYZ+T):this structure corresponds to one video with multiple frames. Each frame can have multiple sections and each section contains one image composed of multiple channels ([1]Video - [1..n]Frame - [1..n]Sections - [1]Image - [1..n] Channels).

Measurements. This level contains the results of the Image and Data Analysis process (cf. Figure 4):

Results of Image Analysis: The results of image analysis are auxiliary images which are usually binary masks or trajectories. These images are results of the application of quality enhancing filters and segmentation algorithms employed to extract regions of

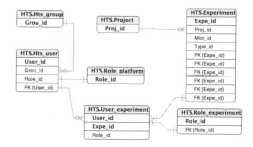

Fig. 5. Database schema for Project Metadata

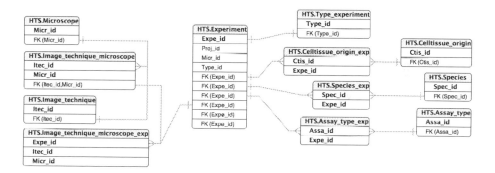

Fig. 6. Database schema for Experiment Metadata

interests (ROIs). These metadata are also linked to the raw video image file, on which the image analysis has been applied.

Results of Data Analysis:
Measurements extracted from the image analysis are further analyzed using pattern recognition tools. After applying operations such as feature selection, clustering and classification, a CSV file is generated with the results accompanied by a HDF file with information of the structure of the CSV file (features).

3 Data Modeling and Storage

The relational database schema designed to store the metadata in an HTS experiment is divided in 5 schema views according to the structure described in Section 2. Figure 5 shows the key components for the project metadata and the possibility to create groups and grant 4 different levels of access to our experiments (Author, Expert User, Analyst User and Guest).

Figure 6 describes the entity Experiment and how the metadata is stored according to the type of experiment performed, the microscope used and the image technique associated. Furthermore, other key components of metadata are mandatory for creating an experiment, such as the specie, assay type and cell/tissue origin.

The most critical part of the metadata corresponds to the Plate-Well metadata shown in figure 7. It requires a validation and verification process before registering new entries

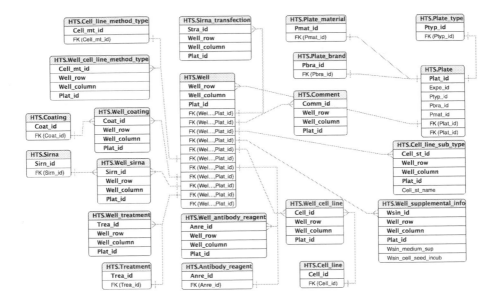

Fig. 7. Database schema for Plate-Well Metadata

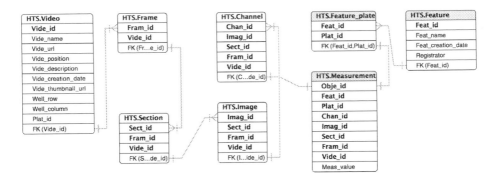

Fig. 8. Database schema for Raw Images and Measurement Metadata

to these entities. The author of an experiment uploads the metadata associated to a plate upon completion of the plate layout design and every entry is validated with the master entities for siRNAs, cell lines, antibodies, reagents, coatings and treatments in order to ensure consistency with the metadata being uploaded.

The main output in an HTS experiment are the raw images or time lapse sequence of images. The image dataset is located on a file server and the URLs to access them are stored in the database (cf. Figure 8). Those images are uploaded to the CytomicsDB through a web interface, which represents the web plate layout interface. During uploading time, using the open source Bio-Formats library [9], a new dataset of images (thumbnails) is generated, and also linked to the raw images in the database. These images are used to have a preview visualization of the plate layout in the web interface.

Another key component to consider in the database schema are the image and data analysis results. The information obtained after the image analysis process is parsed to the database using the entities Features and Measurements (cf. Figure 8). These two entities store the information required by another API such as PRTools [6] to perform the pattern recognition and statistical analysis.

4 Case Study in Endocytosis of EGFR: Identification of EGFR Dynamics and Classification of EGFR Phenotype Stages

In this section we describe a case study on how the structure of the metadata and RDBMS are applied in order to identify the EGFR dynamics and classify the different EGFR phenotypes.

Endocytosis is regarded as a mechanism of attenuating epidermal growth factor receptor (EGFR) signaling and of receptor degradation. Increasingly, evidence becomes available showing that cancer progression is associated with a defect in EGFR endocytosis [5]. Functional genomics technologies combine high-throughput RNA interference with automated fluorescence microscopy imaging and multi-parametric image analysis, thereby enabling detailed insight into complex biological processes, like EGFR endocytosis. The experiments produce over half a million images. Such a volume of images is beyond the capacity of manual processing and therefore, image processing and machine learning are required to provide an automated analysis solution for HTS experiments [2]. The total size in average can vary between 500 Mb to 20 Gb of raw images per experiment and CytomicsDB is designed to cope with the growing data size due to the scalable architecture for storing the images in a File Server and the metadata of the entire experiment in the database.

According to the methodology described in [2], three stages are identified: (1) Image Acquisition, (2) Image Analysis and (3) Data Analysis. We describe each stage as follows:

Image Acquisition: The experiment "Endocytosis of EGFR" is created in the CytomicsDB platform and its respectively plates. The type of metadata required for creating an experiment and the plates in our platform is described in Section 2. The respective values associated to each type of metadata have been detailed in [2]. After designing the plate in the platform, the wet-lab experiment is initiated, which includes the following steps: (1) cell culturing, siRNA transfection and EGF exposure, (2) fluorescent staining of proteins of interest and (3) image acquisition. Upon completion of the acquisition process 960 images are uploaded to the platform which size in total is 767 Mbytes. These images correspond to a 96 wells plate (cf. Figure 9) and for each well, images are captured from ten randomly selected locations. However, an experiment can consist of more than one plate and the number of samples per well can differ per case.

Image Analysis: The API in charge of the image analysis, request from the database the location of each image to process. The query executed is:

```
SELECT v.vide_id, v.vide_name, v.vide_url, v.vide_position, v.well_row, v.well_column
FROM HTS.Video v
WHERE v.plat_id = 17;
```

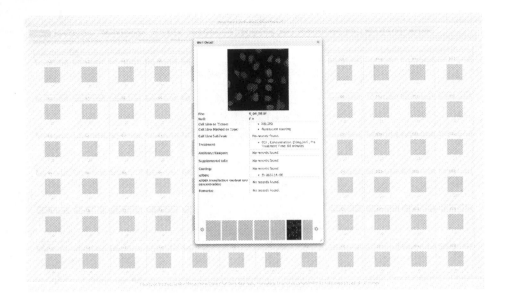

Fig. 9. Web plate layout

The value of column *plat_id* is in this case 17 and it was assigned after selecting *the plate for endocytosis* in the web interface. Three steps are performed by this API: (1) noise suppression, (2) image segmentation and (3) phenotype measurement. The algorithms and process details are described in [2]. Upon completion of the image analysis process, the API returns two outputs: (1) The location in the database of a new set of images and (2) a CSV file containing the features and the phenotype measurement respectively. The set of images generated are: (a) Original image: PERK (red), EFGR (green) and nucleus (blue) (cf. Figure 10), (b) Component definition: artificial cell border (red) and binary mask of protein expression (green) (cf. Figure 11), (c) Cell border reconstruction: artificial cell border (W-V) (cf. Figure 12), (d) Image segmentation: binary mask of EFGR channel by WMC (cf. Figure 13) [13].

The phenotype measurements (CSV file) are parsed first and then stored in the database by a web service executing the following query:

```
INSERT INTO HTS.Measurement
(Obje_id, Feat_id, Plat_id, Chan_id, Imag_id, Sect_id, Fram_id, Vide_id)
VALUES (0,1,17,1,1,1,1,1,14.0);
```

In this example, the column *Feat_id=1* corresponds to *Area* in the entity Feature and the measurement obtained for this feature is 14.0. The column *Plat_id* is still 17 because we refer to the same plate.

The measurements are categorized in two subgroups: (1) basic measurements of the phenotypes covering shape descriptors and (2) the localization phenotype describing the assessment of the correlation between two information channels. The basic phenotype measurement includes a series of shape parameters such as: size, perimeter, extension,

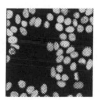 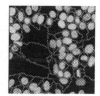 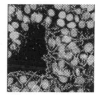

Fig. 10. Original image   **Fig. 11.** Component definition   **Fig. 12.** Cell border reconstruction   **Fig. 13.** Image segmentation

dispersion, elongation, orientation, intensity, circularity, semi-major axis length, semi-minor axis length, closest object distance and in nucleus, these can be extended as the experiments so dictates. In addition to the basic phenotype measurement, localization measurements can be derived for a specific experimental hypothesis. The localization phenotypes are quantifications of comparative measurement between information channels such as relative structure-to-nucleus distance or structure-to-border distance. The features in EGFR-screen based localization phenotypes used are: nucleus distance, border distance and intactness. On the basis of the phenotype measurements, objects are classified into phenotypic stages. For the assessment of significance statistical analysis is performed [2]. Upon completion of the image analysis, it is possible to visualize the results in a web plate layout and export the measurements to files.

Data Analysis: The aim of the endocytosis study is to quantify the process of EGF-induced EGFR endocytosis in human breast cells and to identify proteins that may regulate this process. The EGFR endocytosis process can roughly be divided into three characteristic episodes: i.e. (1) at the onset EGFR is present at the plasma-membrane; (2) subsequently, small vesicles containing EGFR will be formed and transported from the plasma-membrane into the cytoplasm; and (3) finally, vesicles are gradually merging near the nuclear region forming larger structures or clusters. The characteristic episodes are the read-out for HTS. Based on this model it is believed that EGFR endocytosis regulators may be potential drug targets for EGFR-induced breast cancer. Studying each of the stages, i.e. plasma-membrane, vesicle and cluster, may provide a deeper understanding of the EGFR endocytosis process [2].

When the data analysis process is triggered, a web service request to the database (entities feature and measurement) the results from the image analysis process. The output of this web service is the location of a file which contains the results of the test for each siRNA regulator. This file will be requested for the API PRTools for generating classifications and graphs with the comparison of the results, such as: (1) Weighted classification error curve, which represents a combination of a feature selection/extraction method and a classifier algorithm, (2) Results of the feature extraction and (3) Average number of plasma-membrane (a) and vesicle (b) per nucleus [2]. Consolidating in CytomicsDB the experiment's metadata, raw images and images/data analysis results, facilitates further comparison with the result of other HTS experiments.

5 Related Work

In the current area of -omics research, various systems/tools have emerged to try to solve the problem that the existing practice of keeping meta data does not allow for effective data searching and mining. They are generally referred to as Laboratory Information Management System (LIMS).

The work proposed by Colmsee et. al. [4] is probably the closest to CytomicsDB. The authors defined central requirements for a primary lab data management and aspects of best practices to realise those requirements. As a proof of concept, the authors implemented a pipeline to manage primary lab data of crop plants. The pipeline consists of i) data storages including a Hierarchical Storage Management system, an RDBMS and a BFiler package to store primary lab data and their meta information; ii) the Virtual Private Database for the realisation of data security and the LIMS Light application to iii) upload and iv) retrieve stored primary lab data. Compared with this work, CytomicsDB has a more sophisticated data model to cope with different types of data (i.e., images, videos, and data produced in different steps in an HTS experiment), pays special attention to the extensibility of the architecture to enable adding new tools.

In [11], the authors presented three open-source, platform independent software tools for genomic data: a next generation sequencing / microarray LIMS and analysis project center (GNomEx); an application for annotating and programmatically distributing genomic data using the DAS/2 data exchange protocol (GenoPub); and a standalone Java Swing application (GWrap) that provides a GUI for the command line analysis tools. CytomicsDB provides similar functionalities as these tools, but focuses on dealing with Cytomic data. Moreover, for the design of CytomicsDB, we have deliberately chosen for a single integrated system to include all features required for conduction HTS experiments and analysis, instead of individual tools and enabling high profile pattern recognition.

In [12], the authors describe a general modeling framework for laboratory data. The model utilises several abstraction techniques, with focus on the concepts of inheritance and meta-data. In this model, distinct regular entity and event schemas can be defined and fully integrated via a standardized interface. The design allows definition of a processing pipeline as a sequence of events. A layer above the event-oriented schema integrates events into a workflow by defining processing directives, which act as automated project managers of items in the system. This LIMS is built on the Oracle RDBMS, and is maintained by multiple database administrators (DBAs). While with CytomicsDB, our goal is to meet the needs of HTS experiments with a more light-weight, flexible system. By adapting modern web and database technologies, CytomicsDB is easy to maintain (i.e., no DBAs required) and extend (i.e., allowing integrating new tools naturally).

The work by Chan et al. [3] focuses on interactive visualization methods for data generated by HTS experiments. The visualization methods might be adapted by CytomicsDB. However, CytomicsDB is a much more comprehensive information system for HTS data, because it integrates both experiments and analysis data into a single system, and allows various types of users and groups to be defined.

Based on the Golm Plant Database System, Köhl et. al. [7] devised a data management system based on a classical LIMS combined with web-based user interfaces for data entry and retrieval to collect this information in an academic environment. This system stores plant cultivation units in an MS ACCESS database, which would quickly run into scalability issues as the data size grows.

6   Conclusions and Future Work

In this paper, we have presented a semantic approach for organizing metadata and an RDBMS for metadata management in High-Throughput Screening experiments. Our goal is to facilitate the exploration process in the HTS workflow, scientist are aware of semantics and they are pushing forward the need for new approaches in organizing the metadata according to which queries are mostly applied on the data. In HTS, images by itself do not have any meaning, but linking images to their respective metadata allows researchers to learn from their experience and help them in mentalizing semantic structures of the metadata. The RDBMS schema has been designed to support the acquisition, visualization and integration stages using a metadata-based approach. Furthermore, CytomicsDB uses a database engine suitable for applications which demands intensive data mining tasks. Finally, we plan to extend this architecture to a more sophisticated interdisciplinary platform for cytomics. The structure of the metadata proposed in this paper will further evolve to an ontology based framework. A new layer to the architecture will be added in order to perform semantic queries, turning the architecture to a web based interactive semantic platform for cytomics [1].

Acknowledgements. We thank Dr Marjo De Graauw for providing us the data of the experiment included in our case study. This work is partially supported by the Erasmus BAPE program, Cyttron II project (EL), the European FET Flagship Programme the Human Brain Project (www.humanbrainproject.eu/) and the Dutch national project COMMIT (www.commit-nl.nl/).

References

1. Bertens, L.M.F., Slob, J., Verbeek, F.: A generic organ based ontology system, applied to vertebrate heart anatomy, development and physiology. J. Integrative Bioinformatics 8(2) (2011)
2. Cao, L., Yan, K., Winkel, L., de Graauw, M., Verbeek, F.J.: Pattern recogntion in high-content cytomics screens for target discovery - case studies in endocytosis. In: Loog, M., Wessels, L., Reinders, M.J.T., de Ridder, D. (eds.) PRIB 2011. LNCS, vol. 7036, pp. 330–342. Springer, Heidelberg (2011)
3. Chan, T., Malik, P., Singh, R.: An interactive visualization-based approach for high throughput screening information management in drug discovery. In: 28th Annual International Conference of the IEEE Engineering in Medicine and Biology Society, EMBS 2006, pp. 5794–5797 (August 2006)
4. Colmsee, C., Flemming, S., Klapperstuck, M., Lange, M., Scholz, U.: A case study for efficient management of high throughput primary lab data. BMC Research Notes 4(1), 413 (2011)
5. de Graauw, M., Cao, L., Winkel, L., van Miltenburg, M.H.A.M., LeDévédec, S., Klop, M., Yan, K., Pont, C., Rogkoti, V.-M., Tijsma, A., Chaudhuri, A., Lalai, R., Price, L., Verbeek, F., van de Water, B.: Annexin a2 depletion delays egfr endocytic trafficking via cofilin activation and enhances egfr signaling and metastasis formation. In: Oncogene (2013)
6. Duin, R.P.W.: Prtools - version 3.0 - a matlab toolbox for pattern recognition. In: Proc. of SPIE, p. 1331 (2000)
7. Kohl, K., Basler, G., Ludemann, A., Selbig, J., Walther, D.: A plant resource and experiment management system based on the golm plant database as a basic tool for omics research. Plant Methods 4(1), 11 (2008)

8. Larios, E., Zhang, Y., Yan, K., Di, Z., LeDévédec, S., Groffen, F., Verbeek, F.J.: Automation in cytomics: A modern RDBMS based platform for image analysis and management in high-throughput screening experiments. In: He, J., Liu, X., Krupinski, E.A., Xu, G. (eds.) HIS 2012. LNCS, vol. 7231, pp. 76–87. Springer, Heidelberg (2012)
9. Linkert, M., Rueden, C.T., Allan, C., Burel, J., Moore, W., Patterson, A., Loranger, B., Moore, J., Neves, C., MacDonald, D., Tarkowska, A., Sticco, C., Hill, E., Rossner, M., Eliceiri, K.W., Swedlow, J.R.: Metadata matters: access to image data in the real world. The Journal of Cell Biology 189, 1 (2010)
10. Mayr, L., Fuerst, P.: The future of high-throughput screening. Journal of Biomolecular Screening (2008)
11. Nix, D., Sera, T.D., Dalley, B., Milash, B., Cundick, R., Quinn, K., Courdy, S.: Next generation tools for genomic data generation, distribution, and visualization. BMC Bioinformatics 11(1), 455 (2010)
12. Wendl, M., Smith, S., Pohl, C., Dooling, D., Chinwalla, A., Crouse, K., Hepler, T., Leong, S., Carmichael, L., Nhan, M., Oberkfell, B., Mardis, E., Hillier, L., Wilson, R.: Design and implementation of a generalized laboratory data model. BMC Bioinformatics 8(1), 362 (2007)
13. Yan, K., Verbeek, F.J.: Segmentation for high-throughput image analysis: Watershed masked clustering. In: Margaria, T., Steffen, B. (eds.) ISoLA 2012, Part II. LNCS, vol. 7610, pp. 25–41. Springer, Heidelberg (2012)

CUDAGRN: Parallel Speedup of Inferring Large Gene Regulatory Networks from Expression Data Using Random Forest

Seyed Ziaeddin Alborzi[1], D.A.K. Maduranga[1], Rui Fan[1],
Jagath C. Rajapakse[1,2], and Jie Zheng[1,3]

[1] Bioinformatics Research Centre, School of Computer Engineering,
Nanyang Technological University, Singapore 639798
{seyed1,Kasun1}@e.ntu.edu.sg,
{FanRui,ASJagath,ZhengJie}@ntu.edu.sg
[2] Department of Biological Engineering,
Massachusetts Institute of Technology, USA
[3] Genome Institute of Singapore, A*STAR (Agency for Science, Technology,
and Re-search), Bio-polis Street, Singapore 138672

Abstract. Reverse engineering of the Gene Regulatory Networks
(GRNs) from high-throughput gene expression data is one of the most
pressing challenges of computational biology. In this paper a method
for parallelization of the Gene Regulatory Network inference algorithm,
GENIE3, based on GPU by exploiting the compute unified device architecture (CUDA) programming model is designed and implemented.
GENIE3 solves regulatory network prediction by developing tree based
ensemble of Random forests. Our proposed method significantly improves
the computational efficiency of GENIE3 by constructing the forest on the
GPU in parallel. Our experiments on real and synthetic datasets show
that, CUDA implementation outperforms sequential implementation by
achieving a speed-up of 15 times (real data) and 14 to 18 times (synthetic
data) respectively.

Keywords: Gene regulatory network, Random forests, GPU, compute
unified device architecture (CUDA).

1 Introduction

A set of DNA portions which collaborate together and with other objects control RNA and proteins expression levels in a cell is called a Gene Regulatory
Network (GRN). Predicting GRN is critical for perceiving the functioning and
development of biological organisms [1]. Due to progresses in high-throughput
gene expression patterns profiling with DNA microarrays and prevalence of expression data, reverse engineering of GRN from biological data is now widely
used for understanding the underlying mechanisms. However it is still one of
the most challenging tasks in bioinformatics and systems biology. The ability of
GRN models to precisely predict gene expressions would help find interrelated

M. Comin et al. (Eds.): PRIB 2014, LNBI 8626, pp. 85–97, 2014.

genes in a biological process in addition to exploring how a system of genes is influenced by drugs. There are several different methods to predict GRN, including relevance networks [2], empirical Bayesian networks [3], Boolean networks [4,5], Bayesian networks [6,7], and neural network [8]. In spite of intense studies, GRN inference approaches still suffer from low performance. The two main reasons are their incapability of modelling inherent complexities of biological processes and the difficulty to handle high dimensional data (which include expressions of thousands of genes). On account of recent advancement in high-throughput technologies, large datasets are frequently available, thus algorithms and software of high-performance computing for GRN inference with high accuracy is becoming more important for the current research in systems biology.

Within this context, Huynh et al. has applied Random forests to GRN inference in order to tackle the above difficulties [9], because the Random forests method has become popular in handling large datasets as well as high dimensional data [10,11]. Their method, namely GENIE3, was one of the best performers in the DREAM4 in Silico challenge for GRN reverse engineering [12]. Even though it infers GRN with a higher accuracy than other similar methods, it still takes a significant amount of time even for a dataset of moderate size (e.g. less than 50 genes).

In this paper, we present a novel method to accelerate the GENIE3 algorithm based on the model of CUDA programming. In order to increase the speed of GENIE3, for each forest, trees grow in parallel inside GPU. Also, for gaining efficiency, shared memory for fast I/O is exploited. We evaluate our approach for several simulated datasets and one real dataset. Our parallelized approach (named CUDAGRN) is able to achieve a speed-up of 15 times on the real dataset on NVidia Quadra 600 in comparison to the sequential algorithm of GENIE3.

Several methods in computational science and technology have been implemented to run on a GPU in CUDA environment. For instance, GPU implementations have been reported for Smith-Waterman algorithm for sequence alignment [13], robotic multisensory perception [14], structured Bayesian mixture [15], image processing methods [16], mutual information estimation algorithm [17], a PoissonBoltzmann equation solver [18] and biomolecules Del-Phi [19]. To our knowledge, CUDAGRN is one of the first few attempts to parallelize a GRN inference algorithm, which will find applications in many biological problems involving high-throughput data and large regulatory networks.

2   Method

2.1   Sequential Algorithm of GRN Inference

There is an assumption of GENIE3, apart from random noises of the regulatory network, that the gene expression of an individual gene is a function of the expression levels of all other genes. It is assumed that the function defining the expression of gene i can be written in the following formula:

$$Y_j^i = f_i(Y_j^{i*}) + \epsilon_j, \forall_j \in All\ experiments \tag{1}$$

where $Y_j^{i*} = [(Y_j^1, ..., Y_j^{i-1}, Y_j^{i+1}, ..., Y_j^p]$, is the list of input samples, containing values of expression in the jth experiment of all genes excluding gene i and ϵ_j is a random fluctuation with mean of 0. Moreover, GENIE3 algorithm assumes that the function f_i only uses the expression of the genes in Y^{i*} that regulates gene i directly. These are the genes with an edge linked to gene i in the final output network. Constructing regulatory edges connecting to gene i will be finding genes whose expression levels are predictive of the expression of gene i. In terminology of machine learning, the problem is a feature selection problem in regression [20].

Each function of f_i is nonlinear [9] and it has to take into account the expression of a number of genes. Hence, it is required to be fast. Generally, tree-based ensemble approaches, particularly Random forests, are methods of choice to fulfil this purpose. Random forests method is scalable, fast and does not assume the nature of the functions. Also, it can cope with a higher number of features and nonlinear functions [21].

In 2001, Brieman introduced the method of Random forests [22]. From the same dataset, it constructs several decision trees using randomly sampled variables and bootstrapping to generate variant trees to work as an ensemble classifier. In bootstrapping, for each tree new datasets are created uniformly by sampling with replacement cases from the training dataset. Then, these produced bootstraps are used for building trees which are finally aggregated into a forest. It has been demonstrated to be efficacious for datasets which are large and have missing attributes values [22,23]. Two parameters can be configured during the Random Forest training. One is the number of trees which can be adjusted by the user, and the other one is the number of attributes to consider in each split (denoted by K). By tweaking the two parameters, the result can be optimized. Building many decision trees is inefficacious when the trees need to be constructed independently from each other. When the number of trees in the forest is large, a parallel implementation of random forest has the potential to achieve considerable speed-up. On the other hand, it might be an ineffective approach only a small number of trees in the forest.

The GENIE3 algorithm uses the tree-based random forest method to predict a regulatory network. The main idea of the random forest method for inferring a network is to break the problem of constructing a network with p genes into p independent sub-problems. Each sub-problem is defined by a unique learning sample consisting of a pair of input-output sets of the ith gene (denoted by LS) from which the network can be inferred. For instance, the learning sample of gene i is as follows:

$$LS_j^i = (Y_j^{i*}, Y_j^i), j = 1, 2, ..., N \tag{2}$$

where N is total number of samples for each gene, Y_j^{i*} is the set of all samples of input genes and Y_j^i is the set of all samples of output gene i. Taking this learning sample as input, the objective of a GRN inference algorithm is to predict the regulatory links among genes such that it works by first ranking all possible regulatory links from the most significant to the least significant links. Recovery of a network is then achieved by pruning the ranked list of links using a threshold.

In this paper, our focus is mainly on the first one. As such, inference algorithm is introduced here as a process that uses LS to allocate weights to candidate regulatory links from any gene to any other gene, such that edges corresponding to real interactions in the regulatory network would be given higher weights. Each sub-problem which is determined by LS^i, is a regression problem which tries to find a function of f_i to minimize the error in (3):

$$\Sigma_{j=1}^{N} = (Y_j^i - f_i(Y_j^{i*}))^2 \tag{3}$$

In random forest, regression trees [24] solve the above problem. The main idea is to split the learning sample iteratively with binary tests in accordance with one input variable (Y^{i*}) and strive to deduct the output variable variance (Y^i) in the resulting subsets of samples. Candidates are split for variables by comparing with the threshold which is defined as long as the tree grows with the values of input variables. In the method, each tree is constructed on a bootstrap learning sample from the original one, and at each test node, before defining the best split, K attributes are chosen randomly from all attributes which become candidates.

One of the key strengths of the Random forests method is its ability to calculate a variables importance from a tree which allows to rank the input features based on their pertinence for predicting the output [23]. In the Random Forest technique several ways to measure the importance of variables have been recommended. Here, we adopt a measure such that in every test node Z, we can calculate the whole reduction of the variance of the output variable because of the split [25]

$$R(Z) = |S|\sigma^2(S) - |S_t|\sigma^2(S_t) - |S_f|\sigma^2(S_f) \tag{4}$$

where S denotes the set of samples which reach node Z, S_t is its subset for which the test is true, S_f is the subset for which the test is false, $\sigma^2(S)$ is the variance of the output variable in a subset, and $|S|$ denotes the cardinality of a set of samples. For an individual tree, the total importance of one variable is calculated by adding resulting values of nodes in the entire tree where this variable is used to split. Those attributes that are never chosen, receive a value of 0 for their importance, and those attributes that are chosen near the root node generally get high scores. Measures of attribute importance can be extended to ensembles, simply by averaging importance scores over all trees in the ensemble.

The computational complexity of the Random Forests is $O(TKN\log(N))$, where N is the size of the learning sample, K is the number of attributes and T is the number of trees. Therefore, our method has a time complexity of $O(pTKN\log(N))$ as it needs to recover trees in the forest for every p genes. Thus, the computational complexity is log-linear with reference to the number of measurements. In the worst case scenario, it is quadratic in reference to the number of genes since $K = p - 1$. In the next section we will describe how the approach can be parallelized since the p problems, and the generation of T trees, in Random Forest are executed independently from each other.

The final results are directed graphs (i.e. networks) each with p nodes and each node indicates a gene. In each graph, a directed edge from one gene i to another gene j represents that gene i regulates, i.e. represses or activates, the expression of gene j. The objective of inferring the GRN is to find a graph only by analysis of the genes expression in diverse situations. By taking into account the dynamic and combinatorial regulatory relations among genes, the expression levels of individual genes can be predicted. Since the overall procedure of the inference tends to be time-consuming, sometimes taking several days even for datasets of moderate sizes, our goal is to implement parallel versions of GRN inference using CPU cluster and GPU.

2.2 CUDA Programming Model

The general architecture of NVIDIA GPUs with the support of CUDA is illustrated in Figure 1. The GPU has a number of CUDA cores known as shader processors (SP). Each SP has an immense number of registers and a private local memory (LM). Eight SPs together form a streaming multiprocessor (SM). Each SM also contains a particular memory region that is shared among the SPs within the same SM. By combining a number of SMs the GPU is constructed. GPUs also have some additional memories, for instance the global device memory which is accessible from all SPs. The GPU used for the development of our approach and experimental evaluation is the NVidia Quadra 600.

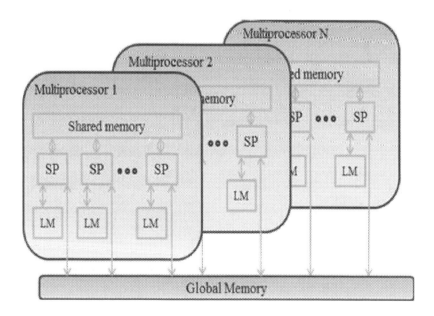

Fig. 1. The GPU architecture assumed by CUDA

The important features of utilized GPU are described in Table 1. For computation in GPU, all data need to be transferred to the GPU memory from the host memory. Therefore the bottleneck of the system is the latency between the CPU and the GPU.

Table 1. The main characteristics for the NVIDIA Quadro 600 graphics card

GPU Property	Values
CUDA cores	98
Compute capability	2.1
GPU / Memory clock rate	1280 Mhz / 800 Mhz
Total amount of memory	1024 MB
Memory interface	128-bit DDR3, 25.6 GB/s

2.3 CUDAGRN Inference Algorithm

All CUDA programs are separated into two sections: (1) sequential codes which are executed on the host CPU and (2) CUDA functions, or kernels, which are launched from the host and executed on the GPU. Before launching a kernel, required data must be transferred to the device memory from the host memory. Since there are several different memories with different sizes accessible by GPU, data transferred into GPU need to be managed, which also can be a bottleneck. In our algorithm, data are arranged based on usage frequency. If data are regularly used, they are moved to the fastest accessible memory, i.e. shared memory. Otherwise, they are stored in the global memory. By placing data in the GPU memory, in a similar way as calling a regular C function, the CUDA kernel is launched. During the execution of a kernel, several CUDA threads are generated and each thread executes an instance of it. Threads are arranged systematically into blocks, and blocks are arranged into grids.

In this paper, we address the problem of parallel constructing Gene Regulatory Networks from gene expression data using the computational power of the GPU. To parallelize the described method in the CUDA environment, some algorithm sections are sent to GPU and executed by GPU threads. The proposed algorithm is highly parallelizable, since all of the p problems of feature selection, solvable by Random Forest, are independent of each other. In addition, different trees in a forest grow independently. Thus, to implement the program in CUDA, forests construction of feature selection problems is achieved in GPU. Because of the memory constraints of device, our approach computes only one problem at a time and for each problem, all of the trees in the forest grow in parallel. As such, it needs a loop of p iterations to accomplish the calculation and come up with the final network. Figure 2 depicts the overall procedure.

As the figure illustrates, we divide the network recovery problem into p isolated sub-problems and iterate p times, where the p is the number of genes, to solve the overall problem. In each loop, constructions of all T trees are parallelized. Furthermore, there is not any straightforward way to recover each single tree in parallel, so we exploit one thread of CUDA to construct a complete tree in the forest. Therefore, our algorithm performs better with a larger number of trees.

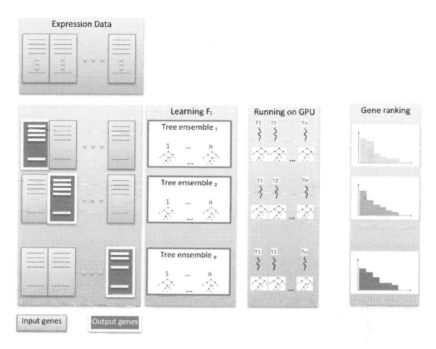

Fig. 2. Parallel procedure for execution of Random Forest algorithm for GRN inference on GPU

Several algorithms for decision tree are developed by recursion. However, using a recursion is not possible for algorithms implemented in CUDA since kernels running on graphic device do not support recursion. Hence, it was necessary to design an algorithm which generates trees iteratively. Algorithm 1 shows the pseudo-code of GENIE3 and algorithm 2 describes the steps in our parallelization of the random forest algorithm for GRN inference.

We have described the details of how trees are built during the training phase. The rest of our approach is similar to the sequential implementation. That is, each tree in the forest is sequentially built by using one thread per tree during the training phase. If N threads are run, then N trees are generated in parallel. Thus, our system works best for an immense number of trees. At each level in a tree, the best attribute to use in order to split a node is picked from a pool of K attributes that are randomly chosen. While all trees are built, they are sent to the host memory for use during the phase of computing variable importance.

Algorithm 1. GENIE3 [9] Pseudo Code

1: **procedure** GRN-INFERENCE SEQUENTIALLY
2:     Data ← *DataReader()*
3:     **For** each sub problem ($i = 1$ to p):
4:         *LearningSamplesGenerator(Data)*
5:         *FeatureSelectionApproach()*
6:         Result ← *CalculateLevelOfConfidence()*
7:         *Normalization*(Result)
8:     **End For**
9:     *RankResult*(Result)
10: **end procedure**

Algorithm 2. CUDAGRN Pseudo Code

1: **procedure** GRN-INFERENCE IN PARALLEL USING GPU
2:     HostMemory ← *DataReader()*
3:     **For** each sub problem ($i = 1$ to p):
4:         *LearningSamplesGenerator()*
5:         DeviceMemory ← *CUDAMemCpy*(HostMemory)
6:         *KernelLaunch()*
7:         **For** each Tree ($j=0$ to NumberOfTreesInForest)
8:             OpenNodeInStack ← *FirstNodeOfATreeGeneration()*
9:             **While**(OpenNodeInStack)
10:                 Node ← OpenNodeInStack[head]
11:                 **If**(*StopSplitFunction*(Node))
12:                     *Leaf(Node)*
13:                 **Else**
14:                     *FindSplit*(Node)
15:                     *Split*(Node)
16:             **End While**
17:             Trees ← *SaveTree()*
18:         **End For**
19:         HostMemory ← *CUDAMemCpy*(Trees)
20:         *GPUMemoryCleanUp()*
21:         Result ← *VariableImportanceCalculator*(*Trees*)
22:         *Normalization*(Result)
23:     **End For**
24:     *RankResult*(Result)
25: **end procedure**

2.4   Parallel GRN Inference on CPU

Message Passing Interface (MPI) is a portable and standardized message-passing system designed to run on a diverse range of parallel computers [26]. Furthermore, OpenMP is an API which supports multi-platform shared memory multi-processing programming on most processor architectures and operating systems [27]. In this section the CPU parallelization of GENIE3, using MPI and OpenMP are presented. Assuming there is enough memory, there are two ways to parallelize GENIE3. In the first way, the algorithm can be parallelized at the tree construction level. For each sub-problem, each CPU thread corresponds to constructing a tree since trees grow independently. Thus, for each sub-problem, a loop with W interactions is executed, where $W = T/thr$, T is the number of trees and thr is the number of active threads. Overall $W*p$ iterations are needed to solve the whole problem. In the second way of parallelizing, each sub-problem must be dealt with in one CPU thread. Therefore, in order to find an answer for a sub-problem, a loop with T iterations is required, and $T*V$, where $V = p/thr$, iterations are required to find a result for all sub-problems.

Since there are overhead costs each time we start OpenMP and MPI, and this can slow down the method, we chose the second way of CPU parallelizing. As such, each CPU thread corresponds to one sub-problem and threads run the problems independently. Eventually, at the end of execution, the main thread accumulates the intermediate results which have been produced by all the threads and provides the final result.

3   Result

In this section, we compare the execution times of the proposed CUDAGRN (described in details in the Methods section) with its sequential and parallel CPU counterparts. The platform for our development was Microsoft Windows 7 along with CUDA version 2.3. Our software also used Core i7 Intel CPU, RAM DDR3 of 8GB as the hardware platform. The GPU was a Quadra 600 NVIDIA with memory of 1GB. Note that we have used a low end GPU versus a high end CPU, which suggests that the observed speedups achieved by CUDAGRN were mainly through the parallelization. Both real and synthetic datasets have been used in our experiments. The real dataset, with 130 experiments and over 6000 genes, was downloaded from http://rana.lbl.gov/EisenData.htm. To further test the scalability of CUDAGRN, we have additionally generated simulated datasets with various parameters, e.g. the numbers of experiments, genes, and trees in the Random forests. Simulated datasets were produced by the software of GeneNetWeaver [28].

Three different versions of the GENIE3 algorithm are implemented and compared in our evaluation, i.e. the sequential C++ program, the CPU-parallelization (by MPI and OpenMP), and the GPU-based version (CUDAGRN). In our experimental evaluation, we studied how the execution time would be influenced by varying different parameters, which include the number of trees to generate (T), the number of genes (p) and sample size. Based on empirical experiments done in [22,23], we configured the number of attributes to sample in each split (K) to its optimal value of $K = \sqrt{p-1}$. However, it is beyond the scope of this paper to prove which configuration of the parameters has the greatest impact on the performance.

Since there are three parameters to vary in the algorithm, we alter one and keep the other two constant. Measurements are collected by running the algorithms on synthetic datasets with the number of trees in a forest from 100 to 10000 and the number of genes from 10 to 500 and sample size (i.e. number of conditions or time points in microarray) from 100 to 5000. Table 2, shows the runtime improvement of CUDAGRN in comparison with other implementations of the GENIE3 algorithm running on different synthetic datasets. As shown in the table, CUDA implementation of the GENIE3 has a faster runtime than the other two implantations when the amount of computation increases. Moreover,

Table 2. Demo result of execution improvement with 1000 trees and 1000 experiments

Number of Genes	Sequential(sec)	CPU 8-Threads(sec)	GPU(sec)
10	30	6	2
20	84	13	6
50	274	38	18
100	753	98	51
200	2175	285	154
500	8361	1093	586

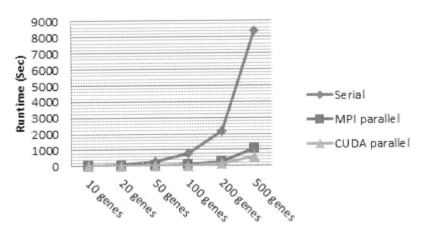

Fig. 3. Diagram of runtimes while the number of genes is varying

line charts of the three tables are shown in Figure 3. The diagram is depicted for a varying number of genes, while the numbers of experiments and trees are both equal to 1000. From the figure, we can see that CUDAGRN achieves the best performance in term of runtime.

In addition, CUDAGRN was faster than the other two implementations when executed on the aforementioned real dataset. CUDAGRN obtained the result of GRN inference approximately 15 times faster than the sequential program (Figure 4). In this section we conducted experiments on real and simulated datasets to show that CUDAGRN is able to infer gene regulatory networks from large and high-dimensional datasets faster than other implementations while maintaining nearly all the accuracy of inference.

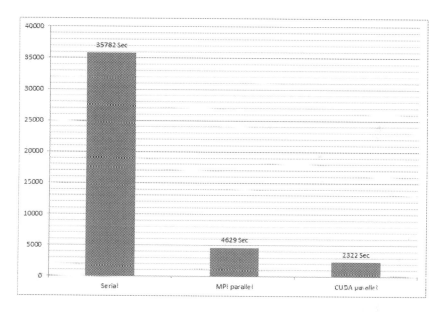

Fig. 4. Computational times of real dataset (6331 Genes, 131 Experiments, 1000 Trees)

4 Conclusion

We presented a novel parallel model of the GENIE3 algorithm, CUDAGRN, developed by exploiting the Compute Unified Device Architecture (CUDA). Comparing the performances of CUDAGRN, Sequential GRN inference and CPU multi-threaded implementations, we observed that CUDAGRN can outperform both the other competitors in term of computational time provided parallelization did not reduce the accuracy of inference.

Unlike the sequential approach and CPU parallelization, the CUDAGRN algorithm as proposed in this paper is executable on the GPUs. On ordinary PCs, the number of processing units (cores) in the CPUs is significantly less than the number of processing units on GPUs. In our approach, whereas the difference in regression performance, e.g. accuracy, among the different implementations is imperceptible (data not shown), it is clear that CUDAGRN is much more efficient in term of computational speed, particularly with large numbers of experiments and trees to build in the Random forest. Testing on real data shows that CUDAGRN is nearly 15 times faster than Sequential GRN, and about twice faster than multi-threaded CPU implementation.

In future, we will refine our implementation of CUDAGRN. In particular, we plan to append properties to make CUDAGRN more accessible to different kinds of applications and practical conditions. For instance, the present version of CUDAGRN is only able to operate on input attributes that are numeric and it cannot deal with missing values which will be addressed in the new version of our implementation.

Acknowledgments. This project has been supported by AcRF Tier 2 grant (ARC9/10, MOE2010-T2-1-056) of Ministry of Education, Singapore.

References

1. Bolouri, H.: Computational modeling of gene regulatory networks: a primer. Imperial College Press, London (2008)
2. Butte, A.J., Kohane, I.S.: Mutual information relevance networks: functional genomic clustering using pairwise entropy measurements. In: Pacific Symposium on Biocomputing, vol. 5, pp. 418–429 (2000)
3. Schafer, J., Strimmer, K.: An empirical Bayes approach to inferring large-scale gene association networks. Bioinformatics 21(6), 754–764 (2005)
4. Liang, S., Fuhrman, S., Somogyi, R.: REVEAL, a general reverse engineering algorithm for inference of genetic network architectures. In: Pacific Symposium on Biocomputing, vol. 3(3), pp. 18–29 (1998)
5. Akutsu, T., Miyano, S., Kuhara, S.: Identification of genetic networks from a small number of gene expression patterns under the Boolean network model. In: Pacific Symposium on Biocomputing, vol. 4, pp. 17–28 (1999)
6. Friedman, N., Linial, M., Nachman, I., Pe'er, D.: Using Bayesian networks to analyze expression data. Journal of Computational Biology 7(3-4), 601–620 (2000)
7. Chen, H., Maduranga, D.A.K., Mundra, P.A., Zheng, J.: Integrating epigenetic prior in dynamic bayesian network for gene regulatory network inference. In: 2013 IEEE Symposium on Computational Intelligence in Bioinformatics and Computational Biology (CIBCB), pp. 76–82 (2013)
8. Vohradsky, J.: Neural model of the genetic network. Journal of Biological Chemistry 276(39), 36168–36173 (2001)
9. Irrthum, A., Wehenkel, L., Geurts, P.: Inferring regulatory networks from expression data using tree-based methods. PloS One 5(9), e12776 (2010)
10. Li, X., Xu, R.: High-dimensional data analysis in cancer research. Springer (2009)
11. Maduranga, D.A.K., Zheng, J., Mundra, P.A., Rajapakse, J.C.: Inferring gene regulatory networks from time-series expressions using random forests ensemble. In: Ngom, A., Formenti, E., Hao, J.-K., Zhao, X.-M., van Laarhoven, T. (eds.) PRIB 2013. LNCS, vol. 7986, pp. 13–22. Springer, Heidelberg (2013)
12. The DREAM4 In Silico network challenge (2010), http://wiki.c2b2.columbia.edu/dream
13. Manavski, S.A., Valle, G.: CUDA compatible GPU cards as efficient hardware accelerators for Smith-Waterman sequence alignment. BMC Bioinformatics 9(suppl. 2), S10 (2008)
14. Ferreira, J.F., Lobo, J., Dias, J.: Bayesian real-time perception algorithms on GPU. Journal of Real-Time Image Processing 6(3), 171–186 (2011)

15. Suchard, M.A., Wang, Q., Chan, C., Frelinger, J., Cron, A., West, M.: Understanding GPU programming for statistical computation: Studies in massively parallel massive mixtures. Journal of Computational and Graphical Statistics 19(2), 419–438 (2010)
16. Park, I.K., Singhal, N., Lee, M.H., Cho, S., Kim, C.W.: Design and performance evaluation of image processing algorithms on GPUs. IEEE Transactions on Parallel and Distributed Systems 22(1), 91–104 (2011)
17. Shi, H., Schmidt, B., Liu, W., Müller-Wittig, W.: Parallel mutual information estimation for inferring gene regulatory networks on GPUs. BMC Research Notes 4(1), 189 (2011)
18. Colmenares, J., Ortiz, J., Rocchia, W.: GPU linear and non-linear Poisson Boltzmann solver module for DelPhi. Bioinformatics, btt699 (2013)
19. Li, L., Li, C., Sarkar, S., Zhang, J., Witham, S., Zhang, Z., Alexov, E.: DelPhi: a comprehensive suite for DelPhi software and associated resources. BMC Biophysics 5(1), 9 (2012)
20. Saeys, Y., Inza, I., Larrañaga, P.: A review of feature selection techniques in bioinformatics. Bioinformatics 23(19), 2507–2517 (2007)
21. Geurts, P., Irrthum, A., Wehenkel, L.: Supervised learning with decision tree-based methods in computational and systems biology. Molecular Biosystems 5(12), 1593–1605 (2009)
22. Breiman, L.: Random forests. Machine Learning 45(1), 5–32 (2001)
23. Breiman, L., Friedman, J., Stone, C.J., Olshen, R.A.: Classification and regression trees. CRC Press (1984)
24. Geurts, P., Ernst, D., Wehenkel, L.: Extremely randomized trees. Machine Learning 63(1), 3–42 (2006)
25. Strobl, C., Boulesteix, A.L., Zeileis, A., Hothorn, T.: Bias in random forest variable importance measures: Illustrations, sources and a solution. BMC Bioinformatics 8(1), 25 (2007)
26. Gropp, W., Lusk, E., Skjellum, A.: Using MPI: portable parallel programming with the message-passing interface, vol. 1. MIT Press (1999)
27. Chapman, B., Jost, G., Van Der Pas, R.: Using OpenMP: portable shared memory parallel programming, vol. 10. MIT Press (2008)
28. Schaffter, T., Marbach, D., Floreano, D.: GeneNetWeaver: in silico benchmark generation and performance profiling of network inference methods. Bioinformatics 27(16), 2263–2270 (2011)

Supervised Selective Kernel Fusion for Membrane Protein Prediction

Alexander Tatarchuk[1], Valentina Sulimova[2], Ivan Torshin[1], Vadim Mottl[1], and David Windridge[3]

[1] Computing Center of the Russian Academy of Sciences, Moscow, Russia
[2] Tula State University, Tula, Russia
[3] Centre for Vision, Speech and Signal Processing,
University of Surrey, Guildford, UK
{aitech,vsulimova,vmottl}@yandex.ru, tiy135@yahoo.com,
D.Windridge@surrey.ac.uk

Abstract. Membrane protein prediction is a significant classification problem, requiring the integration of data derived from different sources such as protein sequences, gene expression, protein interactions etc. A generalized probabilistic approach for combining different data sources via supervised selective kernel fusion was proposed in our previous papers. It includes, as particular cases, SVM, Lasso SVM, Elastic Net SVM and others. In this paper we apply a further instantiation of this approach, the *Supervised Selective Support Kernel SVM* and demonstrate that the proposed approach achieves the top-rank position among the selective kernel fusion variants on benchmark data for membrane protein prediction. The method differs from the previous approaches in that it naturally derives a subset of "support kernels" (analogous to support objects within SVMs), thereby allowing the memory-efficient exclusion of significant numbers of irrelevant kernel matrixes from a decision rule in a manner particularly suited to membrane protein prediction.

Keywords: Multiple Kernel Learning, SVM, supervised selectivity, support kernels, membrane protein prediction.

1 Introduction

Membrane proteins comprise $20 - 30\%$ of all proteins encoded by a genome and perform a variety of functions vital to the survival of organisms. Membrane proteins serve as receptors (i.e. sensors of the cells), transport molecules across the membrane, participate in energy production (ATP biosynthesis) and in cell-cell interaction (cell adhesion) etc. [1]. They are targets of over 50% of all modern medicinal drugs [2]. Consequently, membrane protein prediction, i.e. the classification of proteins as either a membrane or non-membrane is a biomedically important problem, and the subject of much research [3],[4],[5].

This is a typical pattern recognition problem in that the most informative individual feature (in this case typically amino acid sequence data) does not provide the full story. Additional feature information can be derived from a number of

M. Comin et al. (Eds.): PRIB 2014, LNBI 8626, pp. 98–109, 2014.
© Springer International Publishing Switzerland 2014

other sources, such as gene expression data, protein-protein interactions and so on. All these data sources contain different and at least partly independent information about membrane protein prediction task [6]. Consequently, there is a natural desire to incorporate them into a combined prediction rule to decrease prediction-errors.

If the data consisted of vectorized features then this act of combination would constitute a trivial matter of appending feature spaces. However, this is generally not the case for gene-based problems, where data may, for instance, consist only of pairwise comparisons. The most appropriate way for integrating heterogeneous data with a wide variety of gene representations (in this case, amino acid and gene sequences, feature vectors, graphs and so on) thus consists in embedding data objects into representation-specific hypothetical linear spaces via kernel functions and constructing the decision function at the combined space. (A kernel function is any real-valued symmetric function of two-arguments, which forms a semidefinite matrix for any finite collection of objects [7],[8]). In particular, there are a number of approaches in the literature for introducing kernel functions into biomolecular data (cf [7]).

Any kernel function embeds a set of objects into some linear space and plays the role of inner product within it [7],[8]. This fact allows us to employ the kernel-based interpretation of the Support Vector Machine (SVM) method, which was originally designed for linear feature space [9] and is one of the most convenient and effective instruments for the binary classification of objects, forming an optimal linear separating hyperplane from specific "support" training examples.

Mercer Kernels further have the property that linear combinations are also Mercer, meaning that kernel combination is straightforward. There have thus been a number of attempts at combine kernel functions for biomolecular data analysis, the simplest approach being an unweighted sum of kernels. Different linear (or even non-linear) combinations with fixed or heuristically-chosen weights have also been considered; however, overall performance is generally poor.

The most general method of kernel fusion is the approach of Lanckriet et al. [6] which seeks to directly solve for the optimal linear combination of kernels and gives rise to a quadratically-constrained algorithm for determining the non-negative adaptive weights of kernel matrices. The respective kernel combination is incorporated into a decision rule with each kernel's influences on the decision proportional to its weight.

A number of authors have carried this work further in various ways, generalizing the approach to problems other than classification [11],[12], working on algorithmic improvements [13],[14], or deriving theoretical variations, applying different restrictions for weights [15] and making certain theoretical extensions, e.g. weighting not only kernels but also features [16],[17]. These variants typically perform well in constrained scenarios, and where the data are initially represented by feature vectors. However, they tend not to out-perform [6] on real protein data.

Furthermore, most of existing multiple kernel learning methods share a common disadvantage - the absence of a mechanism for supervising so-called "sparseness" of the obtained vector of kernel weights. In the genetic arena, the obtained vector of weights is frequently too sparse, with many informative kernels excluded from the decision rule, with the resulting loss the decision quality.

Only a few methods are explicitly oriented towards elimination of this disadvantage and obtaining non-sparse decisions [19],[20] (more advanced versions utilize a *supervised* sparseness parameter [21],[22],[23],[24]). We refer to this property as "*selectivity*", because it defines an algorithm's ability to select kernels most useful to the classification task at hand. A generalized probabilistic approach for supervised selective kernel fusion was proposed by the authors in [23],[24] and includes, as particular cases, such familiar approaches as the classical SVM [9], Lasso SVM [25], Elastic Net SVM [26] and others.

In this paper we apply a further particular case of this approach, called Supervised Selective Support Kernel SVM (SKSVM), initially proposed in [24] to the membrane protein prediction problem.

We will demonstrate that the proposed approach achieves the top-ranked position among the selective kernel fusion variants on benchmark data set for membrane protein prediction. Uniquely, the proposed approach has the very significant qualitative advantage over the other methods of explicitly indicating a discrete subset of support kernels within the combination, in contrast to the other methods that assign some positive (even if small) weight to *each* kernel, requiring significantly greater memory overhead.

2   Generalized Probabilistic Formulation of the Multiple Kernel Two-Class Recognition Problem

Let $\{(\omega_j, y_j), j = 1, ..., N\}$ be the training set of real-world objects $\omega_j \in \Omega$ (for example, proteins) and $y_j = y(\omega_j) \in \{-1, 1\}$ defines its class-membership. Let also n similarity functions $K_i(\omega', \omega''), \omega', \omega'' \in \Omega, i = 1, ..., n$ be defined, each of which forms a positive semidefinite matrix $\{K_i(\omega_j, \omega_k)\}$ for any finite set of objects $\{\omega_j, \omega_k \in \Omega, j, k = 1, ..., S\}$ and is hence a kernel function [8].

Each kernel function $K_i(\omega', \omega''), i = 1, ..., n$ embeds the set of objects Ω into some hypothetical linear space $\mathbb{X}_i$ by a hypothetical mapping $x_i = x_i(\omega) \in \mathbb{X}_i, \omega \in \Omega$, and plays the role of inner product within it $K_i(\omega', \omega'') = < x_i(\omega'), x_i(\omega'') >: \mathbb{X}_i \times \mathbb{X}_i \to \mathbb{R}$.

For combination using several kernels we here utilize the generalized probabilistic formulation of the SVM, which was proposed in [20,22,23] as an instrument for making Bayesian decisions on the discriminant hyperplane $\sum_{i=1}^n K_i(a_i, \omega) + b \gtrless 0$ within the Cartesian product of the kernel-induced hypothetical linear spaces $\boldsymbol{a} = (a_1, ..., a_n) \in \mathbb{X}, b \in \mathbb{R}$.

The main idea of the proposed probabilistic formulation consists in assuming a specific system of probabilistic assumptions regarding the two distribution densities of hypothetical feature vectors for the two classes: $\varphi(\boldsymbol{x}|y=+1)$ and $\varphi(\boldsymbol{x}|y=-1)$, defined by the (as yet) undetermined hyperplane in the combined linear space $\boldsymbol{x} = (x_1, ..., x_n) \in \mathbb{X} = \mathbb{X}_1 \times ... \times \mathbb{X}_n$ under certain *a priori* probabilistic assumptions.

Let $\boldsymbol{a}^T\boldsymbol{x} + b \gtrless 0$ be some hyperplane with the direction element $\boldsymbol{a} \in \mathbb{X}$ and the bias $b \in \mathbb{R}$. Associated with it are two parametric families of conditional distributions of object densities:

$$\varphi(\boldsymbol{x}|\boldsymbol{a},b,y;c) = const \begin{cases} 1, y(\boldsymbol{a}^T\boldsymbol{x}+b) \geqslant 1, \\ exp\left[-c\left(1-y(\boldsymbol{a}^T\boldsymbol{x}+b)\right)\right], & y\left(\boldsymbol{a}^T\boldsymbol{x}+b\right) < 1. \end{cases} \tag{1}$$

We assume that the random vectors of two classes are distributed substantially within their respective subspaces $\boldsymbol{a}^T\boldsymbol{x} + b > 0$ and $\boldsymbol{a}^T\boldsymbol{x} + b < 0$; the parameter c regulates the extent to which this assumption holds. (Note that fact that the uniform distribution in the upper row of (1) implies an infinite area does not lead to mathematical contradiction, since it participates only in the Bayes' formula[27]).

Suppose the training set $\{(\boldsymbol{x}_j, y_j), \ j=1,...,N\}$, $\boldsymbol{x}_j \in \mathbb{X} = \mathbb{X}_1 \times ... \times \mathbb{X}_n, y_j = \pm 1$ has been obtained. Then the conditional distribution of the whole training set is

$$\Phi(X|Y,\boldsymbol{a},b;c) = \prod_{j=1}^{N} \varphi(\boldsymbol{x}_j|\boldsymbol{a},b,y_j;c). \tag{2}$$

The second key assumption in the proposed probabilistic model is the assumption of a joint *a priori* distribution $\Psi(\boldsymbol{a},b)$ of parameters $(\boldsymbol{a},b)$ defining the separating hyperplane. Assume that we have no any *a priori* preferences about b. We then have that:

$$\Psi(\boldsymbol{a},b) \propto \Psi(\boldsymbol{a}). \tag{3}$$

The *a posteriori* distribution density $P(\boldsymbol{a},b|X,Y;c)$ of parameters $(\boldsymbol{a},b)$ with respect to the training set (X,Y) is then defined by Bayes' formula:

$$P(\boldsymbol{a},b|X,Y;c) = \frac{\Psi(\boldsymbol{a},b)\Phi(X|Y,\boldsymbol{a},b;c)}{const} \propto \Psi(\boldsymbol{a},b)\Phi(X|Y,\boldsymbol{a},b). \tag{4}$$

Understanding the training problem as that of maximizing this *a posteriori* distribution density in the space of model parameters $(\boldsymbol{a},b)$ leads to the criterion:

$$(\hat{\boldsymbol{a}},\hat{b}|X,Y;c) = \underset{\boldsymbol{a}\in\mathbb{X},b\in\mathbb{R}}{\operatorname{argmax}} \left[\ln\Psi(\boldsymbol{a},b) + \ln\Phi(X|Y,\boldsymbol{a},b;c)\right] \tag{5}$$

Theorem 1. *The training criterion (5) for distributional family (1) and a-priori distribution of hyperplane parameters (3) is equivalent to the problem of minimization of the criterion $J(\boldsymbol{a},b,\boldsymbol{\delta}|c)$ in a convex set defined by linear inequality constraints for training objects:*

$$\begin{cases} -\ln\Psi(a_1,...,a_n) + c\sum_{j=1}^{N}\delta_j \to \min\left(a_i\in\mathbb{X}_i, b\in\mathbb{R}, \delta_j\in\mathbb{R}\right), \\ y_j\left(\sum_{i=1}^{n} <a_i, x_i(\omega_j)> +b\right) \geqslant 1-\delta_j, \ \delta_j \geqslant 0, \ j=1,...,N. \end{cases} \tag{6}$$

Kernelizing criterion (6) yields the form:

$$\begin{cases} -\ln \Psi(a_1, ..., a_n) + c \sum_{j=1}^{N} \delta_j \to \min\big(a_i \in \mathbb{X}_i, b \in \mathbb{R}, \delta_j \in \mathbb{R}\big), \\ y_j \left(\sum_{i=1}^{n} K_i(a_i, \omega_j) + b \right) \geqslant 1 - \delta_j, \ \delta_j \geqslant 0, \ j = 1, ..., N. \end{cases} \quad (7)$$

Each specific choice of *a priori* distribution density $\Psi(a_1, ..., a_n)$ expresses a specific *a priori* preference about the hyperplane orientation, and endows training criterion (7) with the ability to select informative kernel-representations (and suppress redundant ones).

In particular, a number of well-known SVM-based training criteria can be obtained form the proposed probabilistic approach, for example, the traditional SVM, Lasso SVM and Elastic Net SVM, differing from one another in the regularization function, which has the form, respectively: $\sum_{i=1}^{n} K_i(a_i, a_i)$, $\sum_{i=1}^{n} \sqrt{K_i(a_i, a_i)}$ and $\sum_{i=1}^{n} K_i(a_i, a_i) + \mu \sum_{i=1}^{n} \sqrt{K_i(a_i, a_i)}$.

3 Supervised Selective Support Kernel SVM (SKSVM)

We apply here a very specific case of the general problem formulation (7), one which was initially proposed in [24]. The *a priori* density of orientation distributions is represented here as composite of the Laplace distribution, while the norms of the components are not less than some given threshold $\sum_{i=1}^{n} \sqrt{K_i(a_i, a_i)} \leq \mu$, and the Gaussian distribution when the norms are greater than the given threshold $\sum_{i=1}^{n} \sqrt{K_i(a_i, a_i)} > \mu$:

$$\begin{aligned} &\psi(a_i|\mu) \propto \exp(-q(a_i|\mu)), \\ &q(a_i|\mu) = \begin{cases} 2\mu \sum_{i=1}^{n} \sqrt{K_i(a_i, a_i)}, & \sum_{i=1}^{n} \sqrt{K_i(a_i, a_i)} \leq \mu, \\ \mu^2 + \sum_{i=1}^{n} K_i(a_i, a_i), & \sum_{i=1}^{n} \sqrt{K_i(a_i, a_i)} > \mu. \end{cases} \end{aligned} \quad (8)$$

The *a priori* assumption of (8) along with the generalized training criterion (7) together define a training optimization problem of the form:

$$\begin{cases} J_{SKSVM}(a_1, ..., a_n, b, \delta_1, ..., \delta_N \,|\, c, \mu) = \\ \sum_{i=1}^{n} q(a_i | \mu) + c \sum_{j=1}^{N} \delta_j \to \min\big(a_i \in \mathbb{X}_i, \ b \in \mathbb{R}, \ \delta_j \in \mathbb{R}\big), \\ q(a_i | \mu) = \begin{cases} 2\mu \sqrt{K_i(a_i, a_i)} & \text{if } \sqrt{K_i(a_i, a_i)} \leqslant \mu, \\ \mu^2 + K_i(a_i, a_i) & \text{if } \sqrt{K_i(a_i, a_i)} > \mu, \end{cases} \\ y_j \left(\sum_{i=1}^{n} K_i(a_i, x_{ij}) + b \right) \geqslant 1 - \delta_j, \delta_j \geqslant 0, j = 1, ..., N. \end{cases} \quad (9)$$

The proposed training criterion (9) is thus a generalized version of the classical SVM that implements the principle of*kernel selection*. We hence refer to the threshold $0 \leqslant \mu < \infty$ in(1) as a "selectivity" parameter because it regulates the ability of the criterion to enact kernel selection. When $\mu = 0 \Rightarrow q(a_i|\mu) = K_i(a_i, a_i)$ the criterion (7) is equivalent to the kernel-based SVM criterion with the minimum ability to kernel selection. At the same time, values

$\mu \gg 0 \Rightarrow q(a_i|\mu) = 2\mu\sqrt{K_i(a_i, a_i)}$ are equivalent to the Lasso SVM with increasing selectivity as μ is increased with respect to the parameter c (until full suppression of all kernels occurs).

Moreover, this criterion, in contrast to other criteria for kernel fusion, explicitly partitions the entire set into two subsets (as is shown in the next section); "support" kernels (which occur in the resulting discriminant hyperplane) and excluded kernels. The proposed approach is hence termed the *Supervised Selective Support Kernel SVM (SKSVM)*.

The approach to solving problem (9) is set out the following two theorems; more detailed description can be found at [24].

Theorem 2. *The decision implicit in problem (9) is equivalent to the decision* $(\hat{\xi}_i \geqslant 0,\ i \in I = \{1, ..., n\},\ \hat{\lambda}_j \geqslant 0, j = 1, ..., N)$ *of the dual problem*

$$
\begin{cases}
L(\lambda_1, ..., \lambda_N | c, \mu) = \sum_{j=1}^{N} \lambda_j - \sum_{i \in I}(1/2)\xi_i \to \max(\lambda_1, ..., \lambda_N), \\
\xi_i \geq 0,\ \xi_i \geq \sum_{j=1}^{N}\sum_{l=1}^{N} y_j y_l K_i(\omega_j, \omega_l)\lambda_j\lambda_l - \mu^2,\ i \in I = \{1, ..., n\}, \\
\sum_{j=1}^{N} y_j \lambda_j = 0,\ 0 \leq \lambda_j \leq (c/2),\ j = 1, ..., N.
\end{cases}
\tag{10}
$$

and is expressed at the form

$$
\begin{cases}
\hat{a}_i = \sum_{j:\hat{\lambda}_j>0} y_j\hat{\lambda}_j x_i(\omega_j),\ i \in I^+ = \left\{ i \in I : \sum_{j=1}^{N}\sum_{l=1}^{N} y_j y_l K_i(\omega_j, \omega_l)\hat{\lambda}_j\hat{\lambda}_l - \mu^2 > 0 \right\}. \\
\hat{a}_i = \hat{\eta}_i \sum_{j:\hat{\lambda}_j>0} y_j\hat{\lambda}_j x_i(\omega_j),\ i \in I^0 = \left\{ i \in I : \sum_{j=1}^{N}\sum_{l=1}^{N} y_j y_l K_i(\omega_j, \omega_l)\hat{\lambda}_j\hat{\lambda}_l - \mu - 0 \right\}, \\
\hat{a}_i = 0,\qquad\qquad i \in I^- = \left\{ i \in I : \sum_{j=1}^{N}\sum_{l=1}^{N} y_j y_l K_i(\omega_j, \omega_l)\hat{\lambda}_j\hat{\lambda}_l - \mu^2 < 0 \right\},
\end{cases}
\tag{11}
$$

4   The Resulting Discriminant Hyperplane and Support Kernels

Assume the dual optimization problem (10) has been solved. Only the Lagrange multipliers $\lambda_1 \geqslant 0, ..., \lambda_N \geqslant 0$ are of interest. In accordance with (11), the solution arrived at partitions the set of all kernels $I = \{1, ..., n\}$ into three subsets:

$$
\begin{aligned}
I^+ &= \left\{ i \in I : \sum_{j=1}^{N}\sum_{l=1}^{N} y_j y_l K_i(x_{ij}, x_{il})\lambda_j\lambda_l > \mu^2 \right\}, \\
I^0 &= \left\{ i \in I : \sum_{j=1}^{N}\sum_{l=1}^{N} y_j y_l K_i(x_{ij}, x_{il})\lambda_j\lambda_l = \mu^2 \right\}, \\
I^- &= \left\{ i \in I : \sum_{j=1}^{N}\sum_{l=1}^{N} y_j y_l K_i(x_{ij}, x_{il})\lambda_j\lambda_l < \mu^2 \right\}.
\end{aligned}
\tag{12}
$$

Theorem 3. *The optimal discriminant hyperplane defined by the solution of the Supervised Selective Support SVM training problem (9) has the form*

$$\sum_{j:\lambda_j>0} y_j\lambda_j \left(\sum_{i\in I^+} K_i(\omega_j,\omega) + \sum_{i\in I^0} \eta_i K_i(\omega_j,\omega) \right) + b \gtrless 0, \qquad (13)$$

where the numerical parameters $\{0\leqslant\eta_i\leqslant 1, i\in I^0; b\}$ are solutions of the linear programming problem:

$$\begin{cases} 2\mu^2\sum\limits_{i\in I^0} \eta_i + c\sum\limits_{j=1}^{n} \delta_j \to \min(\eta_i, i \in I^0; b; \delta_1,\ldots,\delta_N), \\ \sum\limits_{i\in I^0}\left(\sum\limits_{l=1}^{N} y_j y_l K_i(\omega_j,\omega_l)\lambda_l\right)\eta_i + y_j b + \delta_j \geqslant 1 - \sum\limits_{i\in I^+}\sum\limits_{l=1}^{N} y_j y_l K_i(\omega_j,\omega_l)\lambda_l, \\ \delta_j \geqslant 0, \ j=1,\ldots,N, \ \ 0\leqslant\eta_i\leqslant 1, \ i\in I^0. \end{cases} \qquad (14)$$

5   The Subset of Support Kernels

The solution $(\hat{\eta}_i, i \in I^0; \hat{b}; \hat{\delta}_1,\ldots,\hat{\delta}_N)$ of the linear programming problem (14) is completely defined by the training set (X,Y). As is seen from criterion (14), some of coefficients $(\hat{\eta}_i, i\in I^0)$ may equal zero if the respective constraints $0 \leqslant \eta_i$ are active at the solution point.

However, it can be shown that, if all the linear spaces $\mathbb{X}_i$ are finite-dimensional and if the training set is considered as randomly-selected points defined by a continuous probability distribution, then the inequalities $\hat{\eta}_i > 0$ are almost certainly met for all $i\in I^0$.

This means that, without any loss of generality, the constraints $\{0 \leqslant \eta_i \leqslant 1, i \in I^0\}$ may be omitted in (14), and, yet, all kernels $i \in I^0$ will occur in the discriminant hyperplane (13) with nonzero weights. It is hence natural (by analogy with the notion of support objects) to call the subset $I_{supp}=I^+\cup I^0 \subseteq I$ the set of support kernels.

The structure of the subsets of kernels (12) explicitly reveals how the subset of support kernels I_{supp} is affected by the parameter μ in the training criterion (9). Thus, if $\mu = 0$, the set of evident support kernels $I^+\subseteq I$ coincides with the entire set $I=\{1,\ldots,n\}$. In this particular case, the function $q(a_i \mid \mu)$ in (9) is quadratic $q(a_i \mid \mu) = const + K_i(a_i,a_i)$ for all $a_i \in \mathbb{X}_i$, and the training criterion does not differ from the usual SVM without selectivity properties; all of the initial kernels are support kernels because they all occur in the resulting decision rule.

As μ grows, increasing numbers of kernels appear in the set I^- of nonsupport kernels (12), and, correspondingly, the set of support kernels $I_{supp}=I^+\cup I^0$ gets smaller. At the asymptote, the selectivity parameter $\mu \to \infty$ forces all kernels into I^-, such that no support kernels remain at all: $I_{supp} = \varnothing$.

6   Adjusting the Selectivity Parameter

The selectivity parameter $0 \leqslant \mu < \infty$ is thus a structural parameter of the Supervised Selective Support Kernel SVM training criterion that determines a sequence of nested classes of training-set models whose dimensionality diminishes as μ grows, starting from the usual SVM model when $\mu = 0$. As it is not determined *a priori*, at present, the most effective method for choosing the value of the structural parameter is via cross-Validation, directly estimating the generalization performance of the training method.

7   Experimental Design

7.1   Membrane Proteins Data Set

To evaluate the proposed approach as a method for membrane protein prediction we use the same data set as Lanckriet et all. (described in [6]). We thus use as a gold standard the annotations provided by the Munich Information Center for Protein Sequences Comprehensive Yeast Genome Database (CYGD) [28]. The CYGD assigns subcellular locations to 2318 yeast proteins, of which 497 belong to various membrane protein classes. The remaining approximately 4000 yeast proteins have uncertain location and are therefore not used in these experiments.

7.2   Kernel Functions for Membrane Proteins

For the membrane protein prediction we evaluate seven kernel matrices derived from three different types of data: four from the primary protein sequence, two from proteinprotein interaction data, and one from mRNA expression data collected by Lanckriet et all. [6]. (All of these kernel matrices, along with the data from which they were generated are available at *noble.gs.washington.edu/proj/sdp-svm*).

The first two kernel matrices (K_{SW} and K_B) are based on the pairwise sequence alignment algorithms SmithWaterman local alignment (SW) and BLAST (B).

The third kernel (K_{Pfam}) was also derived from protein sequences, but was obtained using hidden Markov models (HMMs) on the Pfam database.

The fourth kernel (K_{FFT}) uses hydropathy profiles, generated from the Kyte-Doolittle index and characterized by alternations of hydrophobic and hydrophilic aminoacids regions which are sufficiently conserved for membrane proteins. The frequency content of the hydropathy profiles, estimated by a FFT procedure, was utilized as a feature vector and used for forming the Gaussian (radial) kernel.

The next two kernels - the linear kernel (K_{Li}) and the diffusion kernel (K_D) are constructed from information about medium- and high-confidence protein-protein interactions from a database of known interactions, which is presented as a matrix [2318x2318] of ones (for pairs of interacted proteins) and zeros (for pairs of non-interacted proteins).

The linear kernel (K_{Li}) matrix is derived from protein feature vectors (i.e. via the inner-product of protein feature pairs).

The diffusion kernel (K_D) considers the interaction-matrix as a graph, in which the nodes corresponded to proteins and the edges to the interactions between them. The diffusion kernel function then measures the similarity of two nodes of the graph based on a randomwalk distance, i.e. such that nodes that are connected by shorter paths (or by many paths) are considered more similar.

Finally, the seventh kernel (K_E) is a radial kernel constructed on the basis of 441-element feature vectors obtained entirely from microarray gene expression measurements. Though gene expression information is not expected to be particularly correlated with any one membrane protein, it is not possible to exclude this kernel *a priori*.

Additionally, five random kernels ($K_{Rnd1}, ..., K_{Rnd5}$) were computed on the basis of 100-length feature vectors, randomly generated without taking into account labeling information about the classes of the proteins. These non-informative kernels were introduced in order to check the ability of the proposed procedure to eliminate non-useful information.

7.3   Experimental Setup

The full set of 2318 proteins (497 membrane proteins and 1821 non-membrane proteins) was randomly split 30 times into training and test sets in the proportion 80:20. As a result, each training set contained 397 membrane proteins and 1456 non-membrane proteins. Each of the test sets contain, respectively, 100 membrane proteins and 365 non-membrane proteins.

For each of 30 training sets obtained we derive 20 different decision rules for membrane protein prediction:

1. For each of 7 informative and 5 random kernels the traditional SVM training procedure was performed separately;
2. SVM classification on the unweighted sum of all 12 kernels was also applied;
3. For all 12 kernels, the proposed Selective Supervised Selective Kernel SVM was performed 6 times with 6 different values of the selectivity-parameter;
4. The optimal decision rule was selected for the proposed method via 5-fold cross-validation.

As a pre-processing step each kernel matrix was centered and normalized to be a unit diagonal matrix.

The quality of each decision was estimated via the ROC-score using the hyperplane bias b to vary sensitivity.

7.4   Results and Discussion

The averages and standard deviations of ROC-scores, computed across 30 randomly generated 80:20 splits for each of 20 training conditions listed in the previous section, are presented in table 1.

As we can see from table 1, the results of the proposed supervised selective support kernel SVM outperform those obtained for each of 12 kernels individually, and also those of the unweighted kernel sum with SVM training. The result obtained at the zero-selectivity level is exactly equal to the result obtained for the unweighted kernel sum (and which supports the theoretical results above).

Table 1. Results of membrane protein prediction

Kernels	algorithm	μ	ROC-score	Kernels	algorithm	μ	ROC-score
K_B	SVM	-	0,825± 0,032	K_{Rnd4}	SVM	-	0,521± 0,029
K_{SW}	SVM	-	0,809± 0,027	K_{Rnd5}	SVM	-	0,509± 0,029
K_{Pham}	SVM	-	0,859± 0,022	All 12	SVM	-	0,881± 0,014
K_{FTT}	SVM	-	0,776± 0,014	All 12	SKSVM	0	0,881± 0,014
K_{Li}	SVM	-	0,634± 0,042	All 12	SKSVM	1	0,881± 0,015
K_D	SVM	-	0,638± 0,037	All 12	SKSVM	5	0,909± 0,014
K_E	SVM	-	0,752± 0,022	All 12	SKSVM	7.5	**0,917± 0,015**
K_{Rnd1}	SVM	-	0,510± 0,029	All 12	SKSVM	10	0,916± 0,015
K_{Rnd2}	SVM	-	0,517± 0,028	All 12	SKSVM	15	0,904± 0,015
K_{Rnd3}	SVM	-	0,515± 0,030	All 12	SKSVM	optimal	**0,918± 0,016**

Moreover, it may be seen that practically all reasonable values of the selectivity-parameter provide good results. The performance obtained using the optimal selectivity value selected via 5-fold cross-validation for each of 30 training sets individually only slightly outperforms the best result obtained using fixed selectivity-levels. This implies that the same selectivity-level is near optimal across the range of training sets (though of course a fixed selectivity level may be not appropriate for different tasks, for example, for recognition different classes of proteins).

The reported results of membrane protein prediction obtained by another multiple kernel learning techniques [6], [16], [18] for the same data set lie in the range [0.87 0.917]. The proposed approach therefore achieves the top-ranked position of the methods reported in the literature. It should be noted, furthermore, that the proposed approach has the unique qualitative advantage of clearly-delineating the subset of support kernels that participate in the decision rule, being thereby directly scientifically interpretable, and potentially assisting with further experimental hypothesis generation.

To demonstrate this delineation of support kernels for one of 30 training sets, table 2 contains the results of the partitionings of the full set of 12 kernels into three subsets: 1) the subset of kernels I^-, which were classified by the algorithm as non-supported, and which are not weighted or included in the decision rule; 2) the subset of kernels I^+ having unit weight and 3) the subset of kernels I^0 having a weight between 0 and 1.

Only the kernels of subsets I^+ and I^0 are support kernels and participate in the decision rule.

As we can see from table 2, the highest selectivity value excludes the random kernels from the set of support kernels entirely. Also, interaction-based linear kernel K_{Li} was excluded in the most cases, while another interaction-based kernel K_D was always excluded.

This result can be explained by such a way. The information about protein-protein interactions can be useful for membrane protein prediction for two reasons. First, hydrophobic molecules or regions of molecules tend to interact with each other. Second, transmembrane proteins are often involved in signaling

Table 2. Kernel fusion results for different selectivity values μ: subsets of non-support (I^-) kernels and support $(I^+$ and $I^0)$ kernels with their weights

μ	K_B	K_{SW}	K_{Pham}	K_{FTT}	K_{Li}	K_D	K_E	K_{Rnd1}	K_{Rnd2}	K_{Rnd3}	K_{Rnd4}	K_{Rnd5}	ROC
0	I^+	I^+	I^+	I^+	I^+	I^+	I^+	I^+	I^+	I^+	I^+	I^+	
	1	1	1	1	1	1	1	1	1	1	1	1	0.877
5	I^+	I^+	I^+	I^+	I^0	I^+	I^+	I^0	I^0	I^0	I^0	I^0	
	1	1	1	1	0.87	1	1	0.26	0.26	0.14	0.36	0.30	0.907
7.5	I^0	I^+	I^0	I^0	I^-	I^+	I^0	I^-	I^-	I^-	I^-	I^-	
	0.70	1.00	0.88	0.81	-	1.00	0.99	-	-	-	-	-	0.919
10	I^0	I^+	I^0	I^0	I^-	I^+	I^0	I^-	I^-	I^-	I^-	I^-	
	0.34	1	0.63	0.56	-	1	0.72	-	-	-	-	-	0.913
15	I^-	I^0	I^0	I^0	I^-	I^0	I^0	I^-	I^-	I^-	I^-	I^-	
	-	0.95	0.14	0.10	-	0.94	0.11	-	-	-	-	-	0.889

pathways, and therefore different membrane proteins are likely to interact with a similar class of molecules upstream and downstream in these pathways. At the same time the diffusion kernel involves the information about interactions more carefully and provides essentially more acurate way for comparing protein sequences in contrast to the linear kernel.

Finally, we can see that only half (6 of 12) of the full kernel set are support kernels in this example, saving on memory requirements.

Thus, in sum, this particular feature of the proposed method makes it preferable to other multi-kernel methods within the literature, which generally assign positive weight to all kernels.

Acknowledgements. We would like to acknowledge support from grants of the Russian Foundation for Basic Research 11-07-00409, 11-07-00728, 14-07-00661, and from UK EPSRC grant EP/F069626/1 (ACASVA).

References

1. Alberts, B., Bray, D., Lewis, J., et al.: Molecular biology of the cell, 3rd edn., p. 1361. Garland Publishing, New York (1994)
2. Overington, J.P., Al-Lazikani, B., Hopkins, A.L.: How many drug targets are there? Nat. Rev. Drug. Discov. 5(12), 993–996 (2006)
3. Krogh, A., Larsson, B., von Heijne, G., Sonnhammer, E.L.L.: Predicting Transmembrane Protein Topology with a Hidden Markov Model: Application to Complete Genomes. J. Mol. Biol. 305, 567–580 (2001)
4. Chen, C.P., Rost, B.: State-of-the-art in Membrane Protein Prediction. Applied Bioinformatics 1, 21–35 (2002)
5. Gao, F.P., Cross, T.A.: Recent developments in membrane-protein structural genomics. Genome Biology 6, 244 (2005)
6. Lanckriet, G., et al.: A statistical framework for genomic data fusion. Bioinformatics 20, 2626–2635 (2004)
7. Schölkopf, B., Tsuda, K., Vert, J.-P. (eds.): Kernel Methods in Computational Biology. MIT Press (2004)
8. Hofmann, T., Schölkopf, B., Smola, A.J.: Kernel methods in machine learning. Ann. Statist. 36(3), 1171–1220 (2008)

9. Vapnik, V.: Statistical Learning Theory. John-Wiley and Sons, Inc. (1998)
10. Pavlidis, P., Weston, J., Cai, J., Grundy, W.N.: Gene functional classification from heterogeneous data. In: Proceedings of the 5th Annual International Conference on Computational Molecular Biology, pp. 242–248 (2001)
11. Ong, C.S., et al.: Learning the kernel with hyperkernels. J. Mach. Learn. Res. 6, 1043–1071 (2005)
12. Bie, T., et al.: Kernel-based data fusion for gene prioritization. Bioinformatics 23, 125–132 (2007)
13. Bach, F.R., et al.: Multiple kernel learning, conic duality, and the SMO algorithm. In: Proceedings of the Twenty-First International Conference on Machine Learning (ICML 2004). Omnipress, Banff (2004)
14. Sonnenburg, S., Rätsch, G., Schäfer, C., Schölkopf, B.: Large Scale Multiple Kernel Learning. Journal of Machine Learning Research 7, 1531–1565 (2006)
15. Hu, M., Chen, Y., Kwok, J.T.-Y.: Building sparse multiple-kernel SVM classifiers. IEEE Transactions on Neural Networks 20(5), 827–839 (2009)
16. Gönen, M., Alpaydın, E.: Multiple Kernel Machines Using Localized Kernels. In: Proc. of PRIB (2009)
17. Gönen, M., Alpaydın, E.: Localized algorithms for multiple kernel learning. Pattern Recognition 46, 795–807 (2013)
18. Liao, L.: Data Fusion with Optimized Block Kernels in LS-SVM for Protein Classification. Engineering 5, 233–236 (2013)
19. Cortes, C., Mohri, M., Rostamizadeh, A.: Learning non-linear combinations of kernels. In: Bengio, Y., et al. (eds.) Advances in Neural Information Processing Systems, vol. 22, pp. 396–404 (2009)
20. Mottl, V., Tatarchuk, A., Sulimova, V., Krasotkina, O., Seredin, O.: Combining pattern recognition modalities at the sensor level via kernel fusion. In: Haindl, M., Kittler, J., Roli, F. (eds.) MCS 2007. LNCS, vol. 4472, pp. 1–12. Springer, Heidelberg (2007)
21. Kloft, M., Brefeld, U., Sonnenburg, S., et al.: Efficient and accurate lp-norm multiple kernel learning. In: Bengio, Y., et al. (eds.) Advances in Neural Information Processing Systems, vol. 22, pp. 997–1005. MIT Press (2009)
22. Tatarchuk, A., Mottl, V., Eliseyev, A., Windridge, D.: Selectivity supervision in combining pattern-recognition modalities by feature- and kernel-selective Support Vector Machines. In: Proc. ICPR (2008)
23. Tatarchuk, A., Sulimova, V., Windridge, D., Mottl, V., Lange, M.: Supervised selective combining pattern recognition modalities and its application to signature verification by fusing on-line and off-line kernels. In: Benediktsson, J.A., Kittler, J., Roli, F. (eds.) MCS 2009. LNCS, vol. 5519, pp. 324–334. Springer, Heidelberg (2009)
24. Tatarchuk, A., Urlov, E., Mottl, V., Windridge, D.: A support kernel machine for supervised selective combining of diverse pattern-recognition modalities. In: El Gayar, N., Kittler, J., Roli, F. (eds.) MCS 2010. LNCS, vol. 5997, pp. 165–174. Springer, Heidelberg (2010)
25. Bradley, P., Mangasarian, O.: Feature selection via concave minimization and support vector machines. In: International Conference on Machine Learning (1998)
26. Wang, L., Zhu, J., Zou, H.: The doubly regularized support vector machine. Statistica Sinica 16, 589–615 (2006)
27. De Groot, M.H.: Optimal Statistical Decisions. Wiley Classics Library (2004)
28. Mewes, H.W., Frishman, D., Gruber, C., Geier, B., Haase, D., Kaps, A., Lemcke, K., Mannhaupt, G., Pfeiffer, F., Schüller, C.: MIPS: a database for genomes and protein sequences. Nucleic Acids Research 28, 37–40 (2000)

Analysis of miRNA Expression Profiles in Breast Cancer Using Biclustering

Antonino Fiannaca[1], Massimo La Rosa[2], Laura La Paglia[1],
Riccardo Rizzo[1], and Alfonso Urso[1]

[1] ICAR-CNR, National Research Council of Italy,
viale delle Scienze Ed.11, 90128 Palermo, Italy
{fiannaca,ricrizzo,urso}@pa.icar.cnr.it, l.lapaglia@libero.it
[2] ICAR-CNR, National Research Council of Italy,
via P. Castellino 111, 80131 Napoli, Italy
larosa@pa.icar.cnr.it

Abstract. RNA sequencing (RNA-seq) is a New Generation Sequencing (NGS) method used for the analysis of transcripts and differential gene expression profiles. With respect to the microarray technology, RNA-seq method has two main advantages: the quantization of a large dynamic range of expression levels, with absolute rather than relative values and no a priori knowledge needed about the sequences to be analyzed. Among small non coding RNAs (sncRNA) obtained through RNA-seq, microRNAs (miRNAs, 22–25 nt long) represent key regulators in multiple cellular functions, as they have a crucial role in different physiological processes. miRNAs are in fact differentially expressed in several types of cancer, in specific tissues and during specific cell status. Since the studies about miRNAs are quite recent, there is not an accurate and unambiguous "pipeline" which can be applied to the study of gene expression by RNA-seq. Clustering algorithms, for instance, have been applied to microarray data in order to discover groups of genes that are co-regulated with respect to certain experimental conditions. Since many regulation mechanisms involve only a set of genes and a limited set of experimental conditions, a new approach is needed. Biclustering is a suitable approach because it can separate groups of rows and columns, in a data matrix, that exhibits similar values or similar characteristics.

In this work we used a biclustering approach in order to identify some patterns of miRNA gene expression deregulation in human breast cancer versus healthy controls. We used the Iterative Signature Algorithm (ISA) tool, which has proved one of the most efficient when applied to gene expression datasets. Considering a real word breast cancer dataset, with our analysis we highlighted 12 miRNA biclusters, each of them involving different types of tumor samples and miRNA families, that were validated in the current scientific literature with the support of the MetaMirClust and UCSC Genome Browser online tools. Our approach has shown the association between specific sub-class of tumor samples having the same immuno-histo-chemical (IHC) and/or histological features. The proposed biclustering methodology has proved a valid instrument for the study of miRNA expression profiles, with the possibility to identify biclusters that

M. Comin et al. (Eds.): PRIB 2014, LNBI 8626, pp. 110–111, 2014.

can provide novel relationships among groups of miRNAs and patient conditions, that eventually have to be validated by in-vitro experiments.

Keywords: Next Generation Sequencing, miRNA expression profiles, breast cancer, biclustering.

Gram-Positive and Gram-Negative Subcellular Localization Using Rotation Forest and Physicochemical-Based Features

Abdollah Dehzangi[1,2], Rhys Heffernan[1], James Lyons[1], Alok Sharma[1,3], Kuldip Paliwal[1], and Abdul Sattar[1,2]

[1] Institute for Integrated and Intelligent Systems (IIIS),
Griffith University, Brisbane, Australia
[2] National ICT Australia (NICTA), Brisbane, Australia
[3] School of Engineering and Physics, University of South Pacific, Fiji
{a.dehzangi,r.heffernan,j.lyons,a.sharma,
k.paliwal,a.sattar}@griffith.edu.au

Abstract. The functioning of a protein relies on its location in the cell. Therefore, predicting protein subcellular localization is an important step towards protein function prediction. Bacterial proteins are considered among the most important proteins that play a wide range of useful and harmful roles. They generally can be divided into two groups namely: Gram-positive and Gram-negative. In this study, we aim at solving Gram-positive and Gram-negative bacterial proteins subcellular localization. Recent studies have shown that using Gene Ontology (GO) for feature extraction can improve the prediction performance. However, for newly sequenced proteins, the GO is not available. Therefore, for these cases, the prediction performance of GO based methods degrade significantly. Besides, the impact of other sources of features such as physicochemical, and evolutionary information to extract local information to tackle this problem, have not been explored adequately.

In this study, we develop a method to effectively employ physicochemical and evolutionary-based information in the protein sequence. To do this, we propose two new segmentation-based feature extraction methods, namely: overlapped segmented density and overlapped segmented autocorrelation. We extract these feature groups by dividing the protein sequence into several segments and extract density and autocorrelation information for each segment in the cumulative, overlapping manner. These feature groups are extracted to capture potential discriminatory information based on physicochemical properties of the amino acids to tackle Gram-positive and Gram-negative subcellular localizations. To incorporate evolutionary information as well, we extract overlapped segmented density and overlapped segmented autocorrelation feature groups from the transformed protein sequence using Position Specific Scoring matrix (PSSM).

We also extract two evolutionary based feature groups, namely: semi composition and auto covariance feature groups. These features groups are extracted to add more evolutionary information to our extracted features. We investigate the effectiveness of our proposed feature extraction

M. Comin et al. (Eds.): PRIB 2014, LNBI 8626, pp. 112–113, 2014.
© Springer International Publishing Switzerland 2014

techniques using 10 attributes. These attributes have been experimentally selected among a wide range of (117 attributes) physicochemical attributes. We also employ Rotation Forest classification technique for our task. Rotation Forest is a newly proposed classifier to build ensemble classifier. It is based on the bagging technique and aims at increasing the diversity in the ensemble classifier. Despite its promising performance for similar studies, it has not been used for protein subcellular localization prediction problem (Figure 1). By applying Rotation Forest to our extracted features, we enhance Gram-positive and Gram-negative subcellular localization accuracies up to 3.4% better than previous studies which also used Gene Ontology for feature extraction.

Keywords: Protein Subcellular Localization, Gram-positive, Gram-negative, Feature Extraction, Physicochemical-based Features, Rotation Forest.

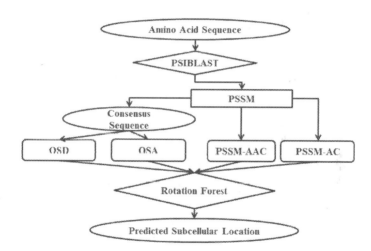

Fig. 1. The overall architecture of our proposed approach

Data Driven Feature Selection
for RNA-Seq Differential Expression Analysis

Henry Han

Department of Computer and Information Science,
Fordham University, New York NY 10023 USA
Quantitative Proteomics Center
Columbia University, New York, NY 10027 USA
xhan9@fordham.edu

Abstract. RNA-Seq provides an unprecedented way to unveil transcriptional details by using ultra-high-throughput sequencing technologies to generate hundreds of million short reads from RNA molecules. As a novel big data, RNA-Seq data challenge existing omics data analytics for its volume and complexity. A major challenge in RNA-Seq analysis is to robustly determine whether an observed count difference of a gene across two or more conditions is statistically significant. Quite a few differential expression (D.E.) analysis models have been proposed to address it from different standing points. They can be categorized as parametric and non-parametric methods according to whether they rely on statistical parameter estimation modeling approaches.

However, almost all these methods do not provide a serious feature selection mechanism but invite all genes into differential expression analysis procedures directly. Such an invite-all scheme not only brings challenges in model-fitting statistically, but also inevitably lowers accuracy in D.E. analysis by increasing false positive ratios. This is because those genes with less contribution to data variations can be falsely detected as D.E. genes by the models, even if count differences for some genes are caused by alignment artifacts, inaccurate library-preparations, or other factors rather than reaction to treatment.

In this study, we presented a data-driven feature selection method: nonnegative singular value approximation to enhance RNA-Seq D.E. analysis by selecting genes with large contributions to data variations along first singular value direction by taking advantages of RNA-Seq count data's built-in characteristics. Unlike most feature selection methods in microarray communities, it does not have any prior data distribution assumption, and avoids any impact from possible distribution modeling biases on D.E. analysis. Moreover, it provides a novel purely data additive way to evaluate each genes significance by avoiding transforming them into a subspace via a linear or nonlinear transform as traditional transform-based feature selection methods such as PCA, ICA or NMF did for microarray data. We further integrated our algorithm with a state-of-the-art D.E. analysis algorithm: DESeq, and found that the proposed algorithm demonstrated advantages in identifying differentially expressed genes with low false positive rates on benchmark data.

M. Comin et al. (Eds.): PRIB 2014, LNBI 8626, pp. 114–115, 2014.
© Springer International Publishing Switzerland 2014

Furthermore, we found that our algorithm contributed to better size factor calculation and more robust parameter estimation in DESeq analysis, in addition to enhancing its sequencing-depth independence in D.E. analysis. Finally, we demonstrated our algorithm's application in RNA-Seq biomarker discovery, which is an important but not addressed topic in RNA-Seq analysis.

Keywords: RNA-Seq, feature selection, differential expression analysis, biomarker.

References

1. Jolliffe, I.: Principal component analysis. Springer, New York (2002)
2. Han, X.: Nonnegative Principal component Analysis for Cancer Molecular Pattern Discovery. IEEE/ACM Transaction of Computational Biology and Bioinformatics 7(3), 537–549 (2010)
3. Robinson, M.D., Oshlack, A.: A scaling normalization method for differential expression analysis of RNA-seq data. Genome. Biol. 11, R25 (2010)
4. Anders, S., Huber, W.: Differential expression analysis for sequence count data. Genome. Biol. 11, R106 (2010)

Intramuscular Fat Percentage Estimation through Ultrasound Images

José Luis Nunes, Alicia Fernandez, and Federico Lecumberry

Facultad de Ingeniería, Universidad de la República
Julio Herrera y Reisig 565, 11300, Montevideo, Uruguay
jlnunes@gmail.com, {alicia,fefo}@fing.com
http://iie.fing.edu.uy

Abstract. This work presents a framework to estimate intramuscular fat percentage on live cattle based on ultrasound images. A procedure to automatically determine the region of interest is proposed. Given the determined ROI, feature extraction and dimensionality reduction is performed based on statistics measures, texture, local binary pattern, among others. A model based in Support Vector Regression (SVR) is trained to estimate the intramuscular fat percentage. A database of ultrasound images acquired by an beef industry expert is used; for each animal there are available the intramuscular fat estimation obtained by an expert using a commercial software, and by chemical analysis. The proposed framework shows good results for a fully automatic procedure.

Keywords: ultrasound images, feature extraction, intramuscular fat estimation, beef quality, support vector regression.

1 Introduction

Beef quality is a complex measure, among others, consumers highlight tenderness as one of the most determinant factors [14,17]. It has been show that intramuscular fat percentage (%IMF) is highly correlated with tenderness [17,2,4]. Therefore an automatic system for its measurement is fundamental.

Intramuscular fat percentage is the proportion of intramuscular fat in the rib eye. This quality measure are usually performed in slaughtered animals. However, is clear the importance and the utility of measuring them with the animal alive, for selective feeding, breeding, rearing [10]. For that reason is becoming important to develop automatic measurements and analysis algorithms on ultrasound images in livestock.

Ultrasound has been used in predicting beef quality for decades, allowing to measure animals' characteristics in a non invasive way and reaching objective measures [5]. It is simple and allows real time evaluations, easy to use in a large group of animals with reasonable costs and offers an alternative for data collection in progeny testing programs.

There are several previous work in this kind of applications, such as [6] addressing the estimation of the %IMF in ultrasound images for livestock. In [3] the rib area was used as a determinant factor in the estimation of beef quality.

M. Comin et al. (Eds.): PRIB 2014, LNBI 8626, pp. 116–122, 2014.

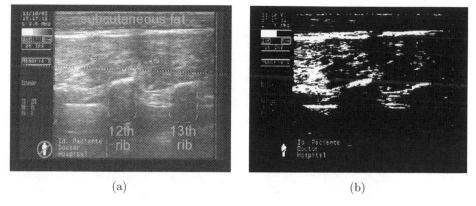

(a)   (b)

Fig. 1. Image examples. Figure 1a shows an example image used as an input for the algorithm. Intramuscular fat appears on the top side, 12th and 13th rib are on the bottom and between them is the ROI. In Figure 1b can see the image in Figure 1a processed by the Otsu's method [15], the result is an binary image where the fat is contained in the white part.

The production method used in Latin America usually includes a high component of extensive farming (although feedlot is used too) impacting in the amount of %IMF, while in other regions the feedlot production is preferred [9,10,1].Therefore, the content of %IMF in animals analyzed in previous works such as [5,10,12,13] might be different from the animals analyzed in the present work and predictors should be adjusted to this case.

In this work we propose an automatic method for feature extraction from the ultrasound image, and adjust a model in order give an estimation of the %IMF. The remainder of the paper is organized as follows section 2 describes the framework, Section 3 presents the experiments and results, and Section 4 gives some conclusions and future work.

2   Framework

The proposed framework performs an automatic procedure for defining the region of interest (ROI) in the ultrasound image (see Figure 1a), and extract a set of features. Then Principal Components Analysis is performed with the extracted features in order to reduce its dimensionality. With this new space of features a Support Vector Regression (epsilon-SVR) was performed to obtain the intramuscular fat estimation. Details of this procedure are given next.

2.1   Defining the Region of Interest Detection

Our interest lies in measuring the %IMF in the *longissimus dorsi* muscle, therefore the ultrasound images are acquired around the 12th and 13th ribs and below the subcutaneous fat [9], only the muscle between the 12th and 13th rib under the subcutaneous fat is taken to determine the %IMF value.

118   J.L. Nunes, A. Fernandez, and F. Lecumberry

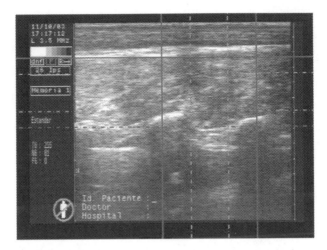

Fig. 2. An image example of the standard output from the automatic ROI detection procedure. Red lines represent subcutaneous fat and both ribs. The square in green is the ROI, an 80×80 pixels square set on the center of the delimited by the subcutaneous fat and the ribs.

In order to automatically determine the region of interest, middle points of both ribs (horizontally) and the upper point were detected, and the lower point of the subcutaneous fat as seen on the Figure 1a. For the subcutaneous fat location the Otsu method [15] was applied for thresholding the image (see Figure 1b). In the binary image we look for a region like the subcutaneous fat. A labeling algorithm was run on the image and then the labeled object with the highest ratio between the horizontal and vertical length was set as the subcutaneous fat. For locating the ribs, an algorithm based on anisotropic diffusion was applied in order to smooth the image without losing border information and restricting small variations of intensity in a same region [16].

Then, a correlation between the image and a synthetic template emulating a generic rib (see Figure 3b) was performed. The two local maxima in magnitude found in the correlation image represent the location of the ribs (see Figure 3a).

Finally, the ROI is defined as a 80×80 pixels square set on the center zone delimited by the subcutaneous fat and the ribs [12]. Figure 2 shows an example of the output from the ROI detection procedure.

2.2 Features Extraction and Selection

A set of forty two features were extracted from each ROI image. These features are based on several statistics and transformation on the ROI, for example, texture descriptors, local binary pattern, co-occurrence matrix, histograms, Fourier Transform, etc. [12,13,11,10].

Features:

Gradient	Co-occurrence matrix
- mean μ $(45°,90°,135°,180°)$	- correlation $(45°,90°,135°,180°)$
- std σ $(45°,90°,135°,180°)$	- homogeneity $(45°,90°,135°,180°)$
Gray Level	- contrast $(45°,90°,135°,180°)$
- mean	- energy $(45°,90°,135°,180°)$
- contrast ratio	
Histogram	**Local Binary Pattern (LBP)**
- percentile (each 20%)	- correlation
- skewness	- homogeneity
Fourier transform	- contrast
- variance coefficient	- energy
- power percentile ($\times$ 5)	

As a result of the feature acquisition stage we obtain a 42-dimension feature space. To reduce the space dimension in order to improve computational performance a feature extraction stage based on Principal Components Analysis was done, finding that 99% of the variance is accumulated in the first ten components. As a result of the PCA a new space of ten new features combinations was used to do the %IMF estimation model.

2.3   Estimation of the %IMF through Support Vector Regression

Support Vector Regression is a variant of the classic Support Vector Machine algorithm. The basic idea of SVR consist in make a mapping of the training data, $x \in \mathbb{X}$, to a larger space $\mathbb{F}$ via a nonlinear mapping $\Phi : \mathbb{X} \to \mathbb{F}$, where a linear regression can be performed. For more details on SVR see [7].

In this work, a radial basis function $(g(u,v) = e^{-\gamma|u-v|^2})$ was used as kernel type. Parameters γ and tolerance of termination criterion were optimized based on the data train set.

3   Experiments and Results

3.1   Database

We worked with a database of 283 ultrasound images (8-bits gray level) obtained from 71 live steers. Four images were taken per animal, and were analyzed independently [12]. Ultrasound images were collected at a cattle ranch in Uruguay. The ultrasound hardware used was the *Aquila Pro Vet*, an industry standard equipment. Based in the ultrasound images an estimation of the %IMF was performed by an expert from the beef industry using a commercial software. Also, the %IMF was measured by chemical analysis and used as *ground truth* to validate the regression results. The lipid extraction protocols used are described in [8]; its margin of error in the measurement is less than 0.3%.

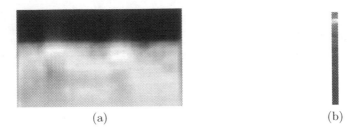

(a)   (b)

Fig. 3. (a) The result of the correlation with the image after the anisotropic diffusion and the template of a synthetic rib, in (b) shows the template who emulates a generic rib

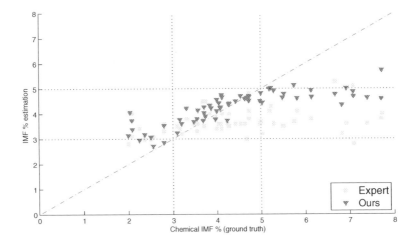

Fig. 4. Scatter plot between both, the %IMF estimation algorithm developed (in purple) and the estimation performed by the expert (in yellow), and the *ground truth* obtained by chemical analysis. The 71 animals are represented in the graphic.

3.2 Performance Analysis

The database was divided into two sets randomly drawn, one to train the algorithm and compute the linear regression coefficients (184 images, 2/3 of the dataset) and the other to test it (92 images, 1/3 of the dataset).

This procedure was repeated 100 times, varying the test and training set. The results were: $RMSE = 1.31$ and $R2 = 0.37$, where $RMSE$ is the root mean square error and $R2$ is the Pearson coefficient of correlation. Figure 4 shows the 71 animals in a scatter plot of the %IMF estimation vs. the *ground truth*. To contrast, the estimation of the %IMF made by the expert, which has an RMSE of 1.58 and a correlation coefficient of 0.23. Table 1 compares the result of the algorithm developed and the expert estimation.

Table 1. Quantitative evaluation of the %IMF estimation algorithm

	Ours	Expert
RMSE	1.31	1.58
R2	0.37	0.23

4   Conclusions and Future Work

A new procedure for estimating the intramuscular fat percentage was presented. First, the region of interest was automatically determined based in ultrasound images characteristics, from this region a set of features was extracted for fat estimation.

The performance of the automatic selection of the ROI was highly satisfactory, more than 96% of the database were well detected, in the reminder 4% where the ROI was wrong detected, the software gives an alert and allows for a manual definition.

The prediction of the intramuscular fat showed a better adjustment in the middle range of fat percentages (3%-5%). Meanwhile for the range of higher fat percentages the error is considerable, underestimating the intramuscular fat, however this error in our approach is lower than the error in the expert's estimation. The overall performance is promising, clearly a deeper analysis of the features considered is needed.

The average execution time is 15 seconds in a standard laptop, which is negligible in terms of industry requirements. Allowing fast estimation of the intramuscular fat percentage at industrial scale.

In future work we propose to study the impact of different parameters in the estimation, such as the ROI's area and location. We also want to explore new textures descriptors in the feature extraction and selection stage.

Acknowledgments. This work was partially supported by ANll grant FMV 2_2011_1_7376. The authors would like to thanks Eileen Armstrong, Gessy Druillet, Marcela Eugster for their contribution on the database acquisition and analysis, Alvaro Gómez, Giovanni Gnemmi, Gregory Randall for their expertise support and Martín Piquerez, Leonardo Pujadas, Alejandro Rivero and Matias Tailanián.

References

1. Aass, L., Fristedt, C.-G., Gresham, J.D.: Ultrasound prediction of intramuscular fat content in lean cattle. Livestock Science 125(2-3), 177–186 (2009)
2. USDA. Agricultural Marketing Services. Usda: Standards for grades of slaughter cattle and standards for grades of carcass beef. Government Printing Office, Washington, D.C. (2006)

3. Arias, P., Sprechmann, P., Pini, A., Sanguinetti, G., Cancela, P., Fernández, A., Gómez, A., Randall, G.: Ultrasound image segmentation with shape periors: Application to automatic cattle rib-eye area estimation. IEEE Transactions on Image Processing 16(6), 1637–1645 (2007)
4. National Cattlemens̉ Beef Association. An evolving industry (2013)
5. Brethour, J.R.: Using receiver operating characteristic analysis to evaluate the accuracy in predicting future quality grade from ultrasound marbling estimates on beef calves. J. Anim. Sci. 78(9), 2263–2268 (2000)
6. Cancela, P., Reyes, F., Rodríguez, P., Randall, G., Fernández, A.: Automatic object detection using shape information in ultrasound images. In: Proceedings of the International Conference on Image Processing, ICIP 2003, September 14-17, vol. 3(2), pp. III – 417–420 (2003)
7. Drucker, H., Burges, C.J.C., Kaufman, L., Smola, A.J., Vapnik, V.: Support vector regression machines. In: NIPS, pp. 155–161 (1996)
8. Folch, J., Lees, M., Sloane Stanley, G.H.: A simple method for the isolation and purification of total lipides from animal tissues. Journal of Biological Chemistry 226(1), 497–509 (1957)
9. Gresham, J.D.: International study guide real time ultrasound beef cattle applications, University Scholars Designation at the Universtiy of Tennesse at Martin (2006)
10. Harron, W., Dony, R.: Predicting quality measures in beef cattle using ultrasound imaging. In: IEEE Symposium on Computational Intelligence for Image Processing, CIIP 2009, pp. 96–103 (March 2009)
11. Jackman, P., Sun, D.-W.: Recent advances in image processing using image texture features for food quality assessment. Trends in Food Science & Technology 29(1), 35–43 (2013)
12. Li, C., Zheng, Y., Kwabena, A.: Prediction of imf percentage of live cattle by using ultrasound technologies with high accuracies. In: WASE International Conference on Information Engineering, ICIE 2009, vol. 2, pp. 474–478 (July 2009)
13. Li, J., Tan, J., Martz, F.A., Heymann, H.: Image texture features as indicators of beef tenderness. Meat Science 53(1), 17–22 (1999)
14. United State Department of Agriculture. Livestock and meat domestic data (December 2013)
15. Otsu, N.: A Threshold Selection Method from Gray-level Histograms. IEEE Transactions on Systems, Man and Cybernetics 9(1), 62–66 (1979)
16. Perona, P., Malik, J.: Scale-space and edge detection using anisotropic diffusion. IEEE Transactions on Pattern Analysis and Machine Intelligence 12, 629–639 (1990)
17. Livestock USDA, Agricultural Marketing Service and Seed Division. United states standards for grades of carcass beef (January 1997)

An Integrated Approach of Gene Expression and DNA-methylation Profiles of WNT Signaling Genes Uncovers Novel Prognostic Markers in Acute Myeloid Leukemia

Erdogan Taskesen[1,2], Frank J.T. Staal[3], and Marcel J.T. Reinders[1,2]

[1] Delft Bioinformatics Lab., Delft University of Technology,
Delft, 2628CD, The Netherlands
[2] Netherlands Bioinformatics Centre (NBIC), The Netherlands
[3] Department of Immunohematology and Bloodtransfusion, Leiden,
University Medical Center, Leiden, 2300 RC, Leiden, The Netherlands
{e.taskesen,m.j.t.Reinders}@tudelft.nl, f.j.t.staal@lumc.nl

Abstract. The development of stem cells and progenitor cells into mature functioning cells depends on, among others, the correct functioning of the wingless-Int (WNT) signaling pathway. WNT signaling has a critical role in regulating stem cell functioning and fate decisions in many different cancerous cells, among them in Leukemia cells by the regulation of haematopoiesis. For Acute Myeloid Leukemia (AML), the malignant counterpart of HSC, currently only a selective number of genes of the WNT pathway are analyzed by using either gene expression or DNA-methylation profiles for the identification of prognostic markers and potential candidate targets for drug therapy. It is known that mRNA expression is controlled by DNA-methylation and that specific patterns can infer the ability to differentiate biological differences, thus a combined analysis using all WNT annotated genes could provide more insight in the WNT signaling.

In this study we had for 344 AML samples, genome wide mRNA expression profiling data (GEP), genome wide DNA-methylation profiling data (DMP), and survival characteristics for the same samples.

We created a computational approach that integrates gene expression and DNA promoter methylation profiles. The approach represents the continuous gene expression and promoter methylation profiles with nine discrete mutually exclusive scenarios. The scenario representation allows for a refinement of patient groups, a more powerful statistical analysis, and the construction of a co-expression network. We focused on 268 WNT annotated signaling genes that are derived from the molecular signature database. Using the scenarios we identified seven independent prognostic markers for overall survival and event-free survival. Three genes are novel prognostic markers; two with favorable outcome and one with unfavorable outcome. The remaining four genes (*LEF1, SFRP2, RUNX1, and AXIN2*) were previously identified but we could refine the patient groups. Three AML risk groups were further analyzed and the co-expression network showed that only the good risk group harbors frequent promoter

M. Comin et al. (Eds.): PRIB 2014, LNBI 8626, pp. 123–124, 2014.

hypermethylation and significantly correlated interactions with protea-
some family members.

Although many studies illustrate that solely gene expression or DNA-
methylation levels can serve as a prognostic marker, our computational
approach proved more effective in the identification of prognostic markers and improves characterization of patient groups by the integration of
gene expression and DNA-methylation profiles. This is especially important when groups are selected on prognostic markers that may be used
for therapeutic interventions.

Keywords: Data integration, Gene expression, DNA-methylation,
Prognostic markers, Acute Myeloid Leukemia, Wingless-Int (WNT).

Improving Performance of the eXtasy Model by Hierarchical Sampling

Dusan Popovic[1,2], Alejandro Sifrim[1,2,3], Jesse Davis[4], Yves Moreau[1,2],
and Bart De Moor[1,2]

[1] KU Leuven, Department of Electrical Engineering (ESAT),
STADIUS Center for Dynamical Systems, Signal Processing and Data Analytics,
Kasteelpark Arenberg 10, 3001 Leuven, Belgium
[2] iMinds Medical IT, Kasteelpark Arenberg 10, 3001 Leuven, Belgium
[3] The Wellcome Trust Sanger Institute, Wellcome Trust Genome Campus,
Hinxton, Cambridge, CB10 1SA, UK
[4] Department of Computer Science, Celestijnenlaan 200A, 3001 Heverlee, Belgium

1 Background

Data from biomedical domains often have implicit hierarchical structure that is usually ignored by practitioners interested in constructing and evaluating predictive models from it. A typical example is genomic data, where a single gene can harbor many mutations while it is at the same time a part of higher-order construct (e.g., chromosome). In such a case different features can be defined over distinct levels of hierarchy. In parallel, a target variable can reflect this intrinsic property of the data, potentially resulting in a biased model.

This happens if the rows (i.e., examples) are inter-dependent, even though the data consists of a single table where each example is described as a fixed-length feature vector. The interdependencies exhibit themselves on different levels of granularity, where all inter-dependent examples have an identical value for a specific feature as well as the same value for the target variable. Thus the feature value appears correlated with the target variable whereas in reality the feature value is correlated with the hierarchical structure of the data. Failing to consider the interdependencies during learning could cause the algorithm to produce a model that simply identifies a pattern that is correlated with the hierarchical structure of the data as opposed to a pattern that is correlated with the target variable.

The described issue figures in the state-of-the-art variant prioritization algorithm called eXtasy[1]. This method incorporates predictors defined over three distinct levels of data granularity - gene level, mutation level and data record level (mutation/phenotype combination), where many data instances share the same values of higher order features (e.g. genes). Here the bias materializes as learning, to a certain degree, to recognize genes which constitute the training set, rather than extracting general characteristics of disease causing mutations. This results in degradation of the performance on the test set.

M. Comin et al. (Eds.): PRIB 2014, LNBI 8626, pp. 125–128, 2014.
© Springer International Publishing Switzerland 2014

2   Methods

We propose a straightforward sampling-based solution for elimination of the described bias. It is implemented and tested within the Random forest framework [2](the eXtasy core model), but it easily generalizes to other types of ensemble learners as well. In particular, instead of extracting a bootstrap[3] from the complete training set to build a single tree on, we first stratify the training examples according to the distinct values of the feature over which the coarsest granularity level is defined. In the case of the eXtasy data, that is the gene identifier. After stratification we randomly select just one data instance from each partition to form the in-bag sample. This prevents algorithm from learning to recognize a particular value of the higher-order feature, as only one example having it will be present in the sample. The procedure differs from the stratification approach that is typically used with the Random forest, where a bootstrap replicate is extracted from each strata to assure that all of them are well-represented.

To ensure a fair comparison, we test the method on the original eXtasy benchmark data using the same evaluation scheme as in the original study. That is, we randomly divide the complete benchmark data set on the gene-level such that two-thirds of the genes belong to the training set and one-third are in the test set. Furthermore, we consider two test scenarios. In the first one we compare two sampling schemes on the unaltered test set, effectively repeating the eXtasy benchmark. In the second one we randomly undersample the positives from the test set in order to mimic the class distributions we would expect to see in the wild; where only one out of 8000 non-synonymous mutation in a genome is potentially disease-causing[4]. We repeat the aforementioned procedure 100 times to stabilize the values of the performance metrics.

3   Results and Conclusions

Under the original eXtasy benchmark scenario, Random forest trained with the hierarchical sampling achieves precision of 0.84, while classical approach results in 0.71. In the same time, the sensitivity drops from 0.86 to 0.78. However, increased Matthews correlation coefficient (from 0.75 to 0.78) indicates that the same sensitivity can be achieved with the hierarchical sampling while maintaining higher precision, by setting an appropriate decision threshold. Furthermore, the realistic class balance scenario underlines this difference even more, as precision doubles (from 0.0024 to 0.0053) with hierarchical sampling, while sensitivity drops from 0.88 to 0.81. In other words, the improved version of eXtasy classifies approximately 188 out of 8000 variants as disease causing, with the probability of capturing the real one equal to 0.81 (i.e. sensitivity). In the same time, the standard eXtasy calls 417 out of 8000 variants, with the probability of hit being 0.88.

Hence, the hierarchical sampling leads to a notable improvement in the model performance in terms of the precision, especially in the most important operating regions for this particular application (i.e. top of prioritized list, see Fig 1.).

Also, as it uses less data (per single tree) than standard Random forests, it results in a more parsimonious model. Finally, we hypothesize that the gain in performance might be even bigger for certain classes of problems. That is, if the number of distinctive values of the coarsest grain concept is much smaller than total number of data records, overfitting on these concepts is more likely to occur. Therefore, such problems could potentially greatly benefit from the proposed sampling scheme.

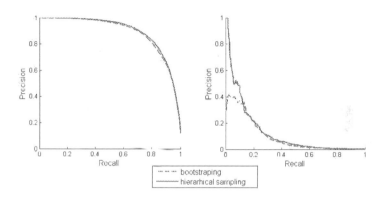

Fig. 1. Precision-recall curves obtained by the application of the eXtasy on the benchmark data (left panel) and the data with the realistic class distribution (right panel). Each panel displays two curves - the one corresponding to the standard Random forest classifier training with bootstrapping and the one corresponding to hierarchical sampling based training.

Acknowledgments. BDM and YM are full professors at the Katholieke Universiteit Leuven, Belgium. Research supported by:

- Research Council KU Leuven: GOA/10/09 MaNet, KUL PFV/10/016 Sym-BioSys, CREA/11/015, OT/11/051, several PhD/postdoc & fellow grants
- Industrial Research fund (IOF): IOF/HB/13/027 Logic Insulin,
  IOF: HB/12/022 Endometriosis
- Flemish Government:
  - FWO: PhD/postdoc grants, projects: G.0871.12N (Neural circuits), research community MLDM
  - IWT: PhD Grants; TBM-Logic Insulin, TBM Rectal Cancer, TBM IETA, O&O ExaScience Life Pharma
  - Hercules Stichting: Hercules III PacBio RS
  - iMinds: SBO 2013; Art&D Instance; ICON b-SLIM
  - IMEC: phd grant
  - VLK Stichting E. van der Schueren: rectal cancer
  - VSC Tier 1: exome sequencing

- Federal Government: FOD: Cancer Plan 2012-2015 KPC-29-023 (prostate)
- COST: Action BM1104: Mass Spectrometry Imaging, Action BM1006: NGS Data analysis network
- EU FP7 Marie Curie Career Integration Grant (294068)

The scientific responsibility is assumed by its authors.

References

1. Sifrim, A., Popovic, D., Tranchevent, L.C., Ardeshirdavani, A., Sakai, R., Konings, P., Vermeesch, J.R., Aerts, J., De Moor, B., Moreau, Y.: Extasy: variant prioritization by genomic data fusion. Nature Methods 10(11), 1083–1084 (2013)
2. Breiman, L.: Random forests. Machine learning 45(1), 5–32 (2001)
3. Efron, B.: Bootstrap methods: another look at the jackknife. In: The Annals of Statistics, pp. 1–26 (1979)
4. Lupski, J.R., Reid, J.G., Gonzaga-Jauregui, C., Rio Deiros, D., Chen, D.C., Nazareth, L., Bainbridge, M., Dinh, H., Jing, C., Wheeler, D.A., et al.: Whole-genome sequencing in a patient with charcot-marie-tooth neuropathy. New England Journal of Medicine 362(13), 1181–1191 (2010)

Ensemble Neural Networks Scoring Functions for Accurate Binding Affinity Prediction of Protein-Ligand Complexes

Hossam M. Ashtawy and Nihar R. Mahapatra*

Department of Electrical and Computer Engineering, Michigan State University
East Lansing, Michigan 48824, U.S.A.
{ashtawy,nrm}@egr.msu.edu

Accurately predicting the binding affinities (BAs) of large sets of protein-ligand complexes (PLCs) is a key challenge in computational biomolecular science, with applications in drug discovery, chemical biology, and structural biology. Since a scoring function (SF) is used to score, rank, and identify drug leads, the fidelity with which it predicts the affinity of a ligand candidate for a protein's binding site has a significant bearing on the accuracy of virtual screening. Despite intense efforts in developing conventional SFs, which are either force-field based, knowledge-based, or empirical, their limited predictive power has been a major roadblock toward cost-effective drug discovery. Therefore, in this work, we present two novel SFs employing a large ensemble of neural networks (NNs) in conjunction with a diverse set of physicochemical and geometrical features characterizing PLCs to predict BA. We build the first ensemble SF, *BgN-Score*, using the bootstrap aggregation technique [1], in which hundreds of neural networks are fitted independently to sets of PLCs sampled randomly without replacement from the training data. BgN-Score calculates the final BA score by computing the average of the predictions of the constituent NNs. The boosting approach [2,3] is used to construct the second ensemble SF, *BsN-Score*, which is based on an additive stagewise fitting of NNs to random sets of PLCs sampled from the training data. The calculated BA of BsN-Score is a weighted sum of the predictions of the individual NNs in the ensemble. These ensemble SFs are trained on 1105 high-quality PLCs retrieved from the 2007 version of PDBbind [4]. Each PLC in this set is characterized using physicochemical features employed by the empirical SFs X-Score [5] (6 features) and AffiScore [6] (30 features) and calculated by GOLD [7] (14 features), and geometrical features used in the machine-learning SF RF-Score [8] (36 features).

We assess the scoring accuracies of the new ensemble NN SFs, BgN-Score and BsN-Score, as well as those of 16 conventional SFs [9] in the context of the 2007 PDBbind benchmark. This dataset encompasses a diverse set of 65 high-quality protein families and contains 195 PLCs independent of the training set. We find that BgN-Score and BsN-Score have more than 25% better Pearson's correlation coefficient (0.807 and 0.815 vs. 0.644) between predicted and measured binding

---

* Corresponding author.

M. Comin et al. (Eds.): PRIB 2014, LNBI 8626, pp. 129–130, 2014.
© Springer International Publishing Switzerland 2014

affinities compared to that achieved by a state-of-the-art conventional SF. In addition, these ensemble NN SFs are also at least 27% more accurate (0.807 and 0.815 vs. 0.631) than SFs based on a single NN that has been traditionally used in drug discovery applications.

References

1. Breiman, L.: Random forests. Machine Learning 45, 5–32 (2001)
2. Friedman, J.H.: Stochastic gradient boosting. Computational Statistics & Data Analysis 38(4), 367–378 (2002)
3. Cao, D.S., Xu, Q.S., Liang, Y.Z., Zhang, L.X., Li, H.D.: The boosting: A new idea of building models. Chemometrics and Intelligent Laboratory Systems 100(1), 1–11 (2010)
4. Wang, R., Fang, X., Lu, Y., Wang, S.: The PDBbind database: Collection of binding affinities for protein-ligand complexes with known three-dimensional structures. Journal of Medicinal Chemistry 47(12), 2977–2980 (2004) PMID: 15163179
5. Wang, R., Lai, L., Wang, S.: Further development and validation of empirical scoring functions for structure-based binding affinity prediction. Journal of Computer-Aided Molecular Design 16, 11–26 (2002) 10.1023/A:1016357811882
6. Zavodszky, M.I., Sanschagrin, P.C., Kuhn, L.A., Korde, R.S.: Distilling the essential features of a protein surface for improving protein-ligand docking, scoring, and virtual screening. Journal of Computer-Aided Molecular Design 16, 883–902 (2002)
7. Jones, G., Willett, P., Glen, R., Leach, A., Taylor, R.: Development and validation of a genetic algorithm for flexible docking. Journal of Molecular Biology 267(3), 727–748 (1997)
8. Ballester, P., Mitchell, J.: A machine learning approach to predicting protein-ligand binding affinity with applications to molecular docking. Bioinformatics 26(9), 1169 (2010)
9. Cheng, T., Li, X., Li, Y., Liu, Z., Wang, R.: Comparative assessment of scoring functions on a diverse test set. Journal of Chemical Information and Modeling 49(4), 1079–1093 (2009)

Integration of Gene Expression and DNA-methylation Profiles Improves Molecular Subtype Classification in Acute Myeloid Leukemia

Erdogan Taskesen[1,2,*], Sepideh Babaei[1,2,*], Marcel J.T. Reinders[1,2], and Jeroen de Ridder[1,2]

[1] Delft Bioinformatics Lab., Delft University of Technology,
Delft, 2628CD, The Netherlands
[2] Netherlands Bioinformatics Centre (NBIC), The Netherlands
{e.taskesen,s.babaei,J.deRidder,M.J.T.Reinders}@tudelft.nl

Abstract. Acute Myeloid Leukemia (AML) is characterized by various cytogenetic and molecular abnormalities. Detection of these abnormalities is important in the risk-classification of patients but requires laborious experimentation. Various studies showed that gene expression profiles (GEP), and the gene signatures derived from GEP, can be used for the prediction of subtypes. Similarly, successful prediction was also achieved by exploiting DNA-methylation profiles (DMP). There are, however, no studies that compared classification accuracy and performance between GEP and DMP, neither are there studies that integrated both types of data to determine whether predictive power can be improved. Here, we used 344 AML samples that can be categorized into 15 well-characterized cytogenetical and molecular subtypes. For each sample, both gene expression and DNA-methylation profiles are available. We created three different classification strategies for the prediction of AML subtypes. For each subtype, we train a two class classifier (logistic regression classifier with lasso regularization) to distinguish between samples that belong to the subtype and samples that belong to the other subtypes. The first classification strategy is no integration of GEP and DMP datasets. The second is early integration by concatenating all GEP and DMP features. The third is late integration based on a two-layer classifier in which the first layer generates optimized sets of parameters for the logistic regression model for separately GEP and DMP, and the second layer employs the posterior probabilities of the GEP and DMP logistic regressors as feature space to train an additional classifier. To assure unbiased measurements of the classification performance of the classifier we followed the double-loop cross-validation protocol. We illustrate that both mRNA expression and DNA-methylation profiles contain distinct patterns that contribute to discriminating AML subtypes and that an integration strategy can exploit these patterns to achieve synergy between both data types. We show that integration of features from both

---

* Shared first authorship.

M. Comin et al. (Eds.): PRIB 2014, LNBI 8626, pp. 131–132, 2014.

data sets improves the predictive power compared to classifiers trained on GEP or DMP alone. In conclusion, we demonstrate that prediction of known cytogenetic and molecular abnormalities in AML can be further improved by integrating GEP and DMP profiles. The best subtype classification accuracy was obtained by the late integration strategy. It outperformed, with one exception, GEP, DMP and early integration for all AML subtypes.

Keywords: Classification, Integration, Gene Expression, DNA-methylation, Acute Myeloid Leukemia subtypes.

The Relative Vertex Clustering Value – A New Criterion for the Fast Discovery of Functional Modules in Protein Interaction Networks

Alioune Ngom, Yifeng Li, and Zina M. Ibrahim

School of Computer Sciences, 5115 Lambton Tower, University of Windsor,
401 Sunset Avenue, Windsor, Ontario, N9B 3P4, Canada

Abstract. Cellular processes are known to be modular and are realized by groups of proteins implicated in common biological functions. Such groups of proteins are called *functional modules*, and many community detection methods have been devised for their discovery from protein interaction networks (PINs) data. In current agglomerative clustering approaches, vertices with just a very few neighbors are often classified as separate clusters, which does not make sense biologically. Also, a major limitation of agglomerative techniques is that their computational efficiency do not scale well to large PINs. Finally, PIN data obtained from large scale experiments generally contain many false positives, and this makes it hard for agglomerative clustering methods to find the correct clusters, since they are known to be sensitive to noisy data.

We propose a local similarity premetric, the *relative vertex clustering value*, as a new criterion allowing to decide when a node can be added to a given node's cluster and which helps addresses the above three issues. Based on this criterion, we introduce a novel and very fast agglomerative clustering technique, FAC-PIN, for discovering functional modules and protein complexes from a PIN data.

Our proposed FAC-PIN algorithm is applied to eight PINs from different species including the yeast PIN, and the identified functional modules are validated using Gene Ontology (GO) annotations. Identified protein complexes are also validated using experimentally verified complexes. Computational results show that FAC-PIN is robust to false positives and can discover functional modules or protein complexes from PINs more accurately and more efficiently than HC-PIN and CNM, the current state-of-the-art approaches for clustering PINs in an agglomerative manner.

M. Comin et al. (Eds.): PRIB 2014, LNBI 8626, p. 133, 2014.

Author Index

Alborzi, Seyed Ziaeddin   85
Ashtawy, Hossam M.   129
Ayadi, Wassim   48

Babaei, Sepideh   131
Belleannée, Catherine   34
Birlutiu, Adriana   10
Bouziri, Hend   48

Cao, L.   72

da Rocha Vicente, Fábio Fernandes   60
Davis, Jesse   125
Dehzangi, Abdollah   112
De Moor, Bart   125
de Ridder, Jeroen   131
Duval, Beatrice   48

Fan, Rui   85
Fernandez, Alicia   116
Fiannaca, Antonino   110

Han, Henry   114
Heffernan, Rhys   112
Heskes, Tom   10

Ibrahim, Zina M.   133

La Paglia, Laura   110
Larios, E.   72
La Rosa, Massimo   110
Lecumberry, Federico   116
Li, Yifeng   133
Lopes, Fabrício Martins   60
Lyons, James   112

Maâtouk, Ons   48
Maduranga, D.A.K.   85

Mahapatra, Nihar R.   129
Marchiori, Elena   23
Moreau, Yves   125
Mottl, Vadim   98

Ngom, Alioune   133
Nicolas, Jacques   34
Nunes, José Luis   116

Paliwal, Kuldip   112
Popovic, Dusan   125

Rajapakse, Jagath C.   85
Reinders, Marcel J.T.   123, 131
Rizzo, Riccardo   110

Sallou, Olivier   34
Sattar, Abdul   112
Sharma, Alok   112
Sifrim, Alejandro   125
Staal, Frank J.T.   123
Sulimova, Valentina   98

Taskesen, Erdogan   123, 131
Tatarchuk, Alexander   98
Torshin, Ivan   98
Tung, Chun-Wei   1

Urso, Alfonso   110

van Laarhoven, Twan   23
Verbeek, F.J.   72

Windridge, David   98

Zhang, Y.   72
Zheng, Jie   85